Mathematics for Retail Buying

4th Edition, Revised

Bette K. Tepper
Fashion Institute of Technology

Newton E. Godnick
Fashion Institute of Technology

Fairchild Publications
New York

First Edition
Copyright © 1973 by Fairchild Publications
Division of Capital Cities Media, Inc.
Second Printing, 1975
Third Printing, 1977

Second Edition
Copyright © 1973 by Fairchild Publications
Division of Capital Cities Media, Inc.
Second Printing, 1980
Third Printing, 1982
Fourth Printing, 1983
Fifth Printing, 1985

Third Edition
Copyright © 1987 by Fairchild Publications
Division of Capital Cities Media, Inc.
Second Printing, 1988
Third Printing, 1990

Fourth Edition
Copyright © 1994 by Fairchild Publications
A Capital Cities/ABC Inc. Company

Fourth Edition, revised
Copyright © 1996
Fairchild Publications, a division of ABC Media, Inc.
Second Printing, 1998

Library of Congress Catalog Card Number: 93-83646
ISBN: 1-56367-088-7

Printed in the United States of America

Contents

UNIT VI   INVOICE MATHEMATICS – TERMS OF SALE   225

UNIT VII   COMPUTER SYSTEM AND CONTROL IN RETAILING   259

List of Figures

Preface

In the first edition of *Mathematics for Retail Buying*, we stated that operating figures are the language of the retail amerchandiser, regardless of the size of the store. In today's highly competitive business environment, it is more important than ever to have comprehension of the mathematical factors involved in profitable merchandising. Without this background, it is difficult to understand the operations of either a small store or a department in a large store. This fundamental mathematical background gives insight into how merchandising problems are solved mathematically and perception as to why merchandising decisions are based on figures. The most skilled and experienced merchandiser knows that the mastery of mathematical techniques and figure analysis is an essential tool. Additionally, persons in related careers who comprehend the mathematics of profit factors will also benefit by their broader understanding of merchandising situations.

Mathematics for Retail Buying delineates the essential concepts, practices, and procedures as well as the calculations and interpretations of figures related to the many factors that produce profit. The choice of material and the depth of each subject are deliberately confined to that which has practical value for the performance of occupations in or associated with retail merchandising. The book was developed primarily as a learning device to meet the needs of career-oriented students who will be directly, or indirectly, involved in the activities of merchandising and buying at the retail level. In addition to the obvious educational benefits to the students of retailing, *Mathematics for Retail Buying* can serve as a guide in training junior executives, and can be a constant source of reference for an assistant buyer or the merchant who operates a small independent store.

When contemplating the writing or revision of any retailing textbook, the question of practical value inevitably will arise. We feel the contents of *Mathematics for Retail Buying* are not only sensible, but realistic because it has been based on:

▶ The personal experience of both authors, who were buyers in major retail organizations.

▶ The knowledgable opinions of current retail executives as to the information necessary to prepare prospective retail merchandisers.

▶ The insights manufacturers have gained that helped to provide better service and to communicate more effectively with retailers and merchandisers.

▶ The tested use of the material by the authors and other teachers at the Fashion Institute of Technology, by faculties at other schools, and by retail executives and training departments.

▶ The favorable comments about the value of the material in relation to career performance by the alumni of the Fashion Buying and Merchandising curriculum at the Fashion Institute of Technology.

The study and comprehension of the principles and techniques contained in *Mathematics for Retail Buying* will enable the student to:

▶ Recognize the basic and elementary factors of the buying and selling process that affect profit.

▶ Understand the relationship of the profit factors and improve profit performance by the manipulation of these factors.

- ▶ Become familiar with the use and function of typical merchandising forms encountered in stores.
- ▶ Become aware of practices and procedures in stores.
- ▶ Become familiar with the applications of computers and data processing in retailing.
- ▶ Understand and apply the basic mathematical concepts used to solve real-life merchandising problems.
- ▶ Comprehend the standard industry terminology employed in retailing and merchandising.

Every effort has been made to build the contents of the book into a solid foundation for ease of understanding, and not merely to display formulas. Without sacrificing completeness, the material has been structured into a simplified outline form. This step-by-step presentation uses brief explanations stated in basic terms, precise definitions, typical merchandising forms illustrating procedures and techniques, and concepts demonstrated clearly in examples that express the mathematical principles involved. An outstanding feature of the book is that it has the combined advantages of a handbook and the benefits of a workbook. Within each unit, after a major topic, a series of problems is presented that tests the understanding of the fundamental principles discussed in that section. Contained at the end of each unit, there is a large collection of review problems and a case study that utilizes practical retailing situations and common difficulties and obstacles encountered in real-life merchandising. These two features will further reinforce comprehension of that specific unit, and consequently, the book.

For this fourth edition, revised, particular attention has been paid to the comments of the various users, including our own teaching experiences. The significant changes in this edition are:

- ▶ The new Unit III, "The Relationship of Markup to Profit" combines the averaging of markup calculations with the various types of markup (i.e., cumulative, initial, and maintained) that pertain to and have a direct effect on profit.
- ▶ The new Unit VII, "Computer System and Control in Retailing" discusses the computer's basic functions, industry-related strategies, and common merchandising and operational information provided by computerized merchandising systems.
- ▶ The addition of a comprehensive instructor's guide for the entire text, which also includes the computations and answers to the practice problems and case studies for each unit.
- ▶ At the end of Units I-VI, a case study problem has been added to demonstrate the use of that material in a realistic, practical merchandising situation.
- ▶ The calculation of the gross margin return on inventory (GMROI) is explored and demonstrated as the measurement of investment efficiency.
- ▶ The basic stock method is added to the other approaches of balancing stock and sales.
- ▶ Up-to-date computer store forms present current merchandising procedures and techniques.
- ▶ The text material concerning the basic principles of pricing and repricing merchandise has been expanded and separated from the section on averaging markup technique.

- The calculation of the cost of merchandise sold that determines the all-important gross margin figure reinforces the existing retail method of inventory material.
- Current industry terminology and data is reflected in the text and problems presented.
- Additional concepts and calculations that illustrate net sales, expenses, and the contribution method are included in the profit and loss section.
- The addition of new, up-to-date review problems.
- The previous sequence of the units has been altered.
- An index is provided to facilitate the student's ease of use of the text and understanding of terminology.
- To assist the student's performance, selected answers are given at the end of the book.
- Additionally, this fourth editiion has now been revised and further improved. We have corrected the reported errata from the first printing, as well as clarifying some salient issues and points. We have also condensed the spacing of the practice problems, in addition to eliminating excess notes pages and blanks, in an effort to make the book slimmer and easier to handle.

In developing and changing the book for this fourth edition, revised, we have tried to be sensitive to and concerned about comments from colleagues and students at the Fashion Institute of Technology and from many other colleges relating to the sequence of study units in our book. The old puzzle about "the chicken or the egg" has surfaced time and time again. Why the Profit unit first? Why not the Retail Method unit? Or Dollar Planning? Or Pricing?

We pondered this matter rather endlessly before we attempted the third revision a few years ago. At that time, we happened upon a rather basic principle about the buying/merchandising process: there is no one beginning and no one end. It is a circular process: an unending, interrelated continuum within which all factors influence all other factors. Even though the units are interrelated, each chapter can be used as a separate topic of study, and each can be shifted without losing the continuity of the process. The various facets of the merchandising cycle are an ongoing procedure of monitoring, analysis, and evaluation directed to the purpose of meeting the merchandiser's objective of achieving a pre-determined gross margin which should, in turn, generate an appropriate net profit. This is mentioned here to point out to instructors and students alike (and to merchants who use our book for in-service training purposes) that the study of retail mathematics may begin at any point on the "circle."

In this fourth edition, revised, the material is divided into seven units. Each unit covers a particular, basic mathematical factor that affects the profits of a retail store. The relationships among these profit factors are stressed throughout the book. The units of study and their subject matter are:

- Unit I, Merchandising for a Profit — introduces the concept of profit, and presents the calculation, interpretation, and analysis of the profit and loss statement.
- Unit II, Retail Pricing and Repricing of Merchandise — discusses and illustrates the basic pricing factors used in buying decisions and presents the calculations used when pricing and/or repricing retail merchandise.
- Unit III, The Relationship of Markup to Profit — explains the

importance of markup to profitable merchandising, and illustrates the calculations of the various types of markup that, when understood and implemented, achieve the desired results.

▶ Unit IV, The Retail Method of Inventory — presents and explains this proven, reliable procedure as a mechanism and system of determining the total value of the stock-on-hand and shortages.

▶ Unit V, Dollar Planning and Control — includes the analysis of possible sales, the planning and control of stocks and purchases, and the techniques used to accomplish these objectives.

▶ Unit VI, Invoice Mathematics — Terms of Sale — is concerned with discounts, dating procedures, and shipping terms that a retail buyer must know to buy effectively and profitably at wholesale.

▶ Unit VII, Computer Systems and Control in Retailing — introduces the standard industry terminology and analyzes the fundamental functions performed by computers in retailing by presenting the available industry-wide merchandising strategies. Additionally, reports that furnish periodic merchandising and operating information are illustrated.

The idea for this book originated with the many faculty members who helped develop some of this material to meet the needs for teaching merchandising mathematics at the Fashion Institute of Technology. The authors are indebted to those faculty members for their contributions and to the many students and alumni for their constructive criticism and suggestions that were used in preparing this book. The enthusiastic response provided the incentive to complete this revision.

We would like to thank the following people for their constructive, helpful, and thoughtful reviews and critiques of the revision in all of its various forms and states: Lester Asher, Fashion Institute of Technology; Suzanne Coil, Baker University; Doris H. Kincade, Ph.D., Virginia Polytechnic Institute & State University; Grace I. Kunz, Ph.D., Iowa State University; Barbara A. Little, The Bradford School; Judith Lusk, Ph.D., C.H.E., Baylor University; Dr. Judy K. Miler, University of Tennessee at Chattanooga; Gwendolyn S. O'Neal, Ph.D., The Ohio State University; Kathleen A. Peters, Arizona State University at Tempe; Aaron Schorr, Fashion Institute of Technology; Carol F. Tuntland, California State University Los Angeles; Andrea L. Weeks, Business & Systems Analyst, Strouds Linens.

We want to acknowledge and give special thanks to the following people for their valuable assistance in the revision of this edition: Katherine Mack, Manager of Executive Training at Bloomingdale's, for the contribution of current forms and reports that illustrate the use of computers in retailing; Maria Cosentino, Divisional Merchandise Manager of A & E Stores, Inc., for sharing her past and present experiences and knowledge in the buyer's use of computers in retailing; Ronald Grille, Vice President, Richter Management Services, Inc., for the examples of computer programs available to retailers; Robert L. Benjamin, editor, for his patience, fortitude, and good humor in checking all the calculations; Jack Tepper, former Vice President Divisional Merchandise Manager of Ohrbach's, and International Consultant to Mondial International, to whom we are indebted for providing the criticism, advice, guidance, inspiration, and steadiness in presenting practical, merchandising situations in the text, problems, and case studies; and Susan Jeffers Casel, editor, for her support, patience, diligence, and professionalism.

Bette K. Tepper
Newton E. Godnick

Mathematics
for Retail Buying

UNIT I
▶ Merchandising for a Profit

▶ Recognition of the importance of profit calculations in merchandising decisions.

▶ Identification of components of a profit/loss statement.

▶ Completion of a profit/loss statement, including the calculation of:
Net sales
Cost of goods sold
Gross margin
Operating expenses
Profit

▶ Identification of types of business expenses and their impact on profit.

▶ Utilization of profit calculations to:
Make comparisons between departments and/or stores
Detect trends
Make changes in merchandising strategy to achieve an increase in profits

KEY TERMS

alteration and workroom costs

billed cost

cash discounts

contribution

controllable expenses

cost of merchandise sold

customer allowances

customer returns

direct expenses

gross margin

gross profit

gross sales

indirect expenses

inward freight

Merchandising and Operating Results of Department and Specialty Stores

net sales

noncontrollable expenses

net profit

profit/loss statement

total merchandise handled

hy is the study of the calculation of profit necessary? An individual involved in retailing inevitably will have many opportunities to become involved with the concept of profit. An employee of a private organization should be aware of corporate profits. For example, many companies offer a profit-sharing plan, a form of incentive common today in many industries. An individual may be a possible investor of personal funds in publicly-owned corporations. To an employee of a publicly-owned corporation, profit is a most significant goal in the sale of shares. Since the beginning of the 20th century, the United States government has required all entrepreneurs to declare the profits or losses of all business ventures. The profits are taxed. Competent accounting methods require a statement of net profit before and after taxes.

One of the major responsibilities for a merchandiser in retailing will be to attain a profit for the department[1], store, or retail business being supervised. For the merchandising executive, the calculation of profits is used to:

▶ Exchange data and compare stores to determine strengths and weaknesses.

▶ Indicate the direction of the business and whether it is prosperous, struggling for survival, or bankrupt.

▶ Provide a statement for analysis so that knowledgeable changes in management or policy can be made.

▶ Improve the profit margin by using this analysis.

In this unit, the basic profit elements, their relationship to each other in skeletal as well as final profit and loss statements, and methods to manipulate these elements to produce improved profits will be examined. When the study of these topics is completed, the meanings understood, the calculations learned, and the existing inter-relationships discerned, the analytic determination of whether an organization is operating at a profit or a loss is possible.

▶ I. DEFINING THE BASIC PROFIT FACTORS

The three basic merchandising factors that directly affect profit results are: OPERATING INCOME (retail stores refer to this as sales volume, which is a net sales figure), COST OF MERCHANDISE SOLD, and OPERATING EXPENSES.

The function of the retail store is to sell merchandise to consumers at a profit. These purchases are the store's source of operating income, known as the store's SALES VOLUME. This is a NET SALES figure.

RETAIL is the price that stores offer merchandise for sale to the consumer. GROSS SALES are the total of all the retail prices charged to consumers on individual items of merchandise, multiplied by the number of units actually sold, or the entire dollar amount received for merchandise sold during a given period. Stores typically give customers the privilege of returning merchandise, and when merchandise is returned to stock and the customer is given money back or is given a credit, these returns of sales are called CUSTOMER RETURNS. Additionally, if after a sale has been made the customer is then given a reduction in price, this is known as a CUSTOMER ALLOWANCE. NET SALES is the sales total for any given period after both customer returns and

---

[1]Department — merchandise department: A grouping of related merchandise for which separate expense and merchandising records are kept, for the purpose of determining the profit of this grouping. It is not merely a physical segregation.

allowances have been deducted from GROSS SALES. When retailers compute profit, the net sales figure is the more significant, because a firm can only realize a profit on goods that remain sold at the retail prices charged.

Merchandisers must determine the retail price for the items purchased, and they must also be concerned with how much they can afford to pay a vendor for merchandise, which is referred to as the COST. TOTAL COST OF MERCHANDISE OR GOODS SOLD or COST OF SALES is the invoice cost of purchases with the necessary adjustments of BILLED COST, TRANSPORTATION COSTS, ALTERATION AND WORKROOM COSTS, and CASH DISCOUNTS taken into account.

The retailer must maintain a place of business from which the goods are sold, and to maintain this place of business, operating expenses must be incurred. Operating expenses commonly fall into two major categories and are charged to a merchandise department to determine the net profit for that department.

Expenses that exist only with a given department and cease if that department is discontinued are called DIRECT EXPENSES. These expenses include salaries of the buyer, assistant buyer, and salespeople; departmental advertising, selling supplies, and customer delivery expenses. Store expenses that exist whether a department is added or discontinued are INDIRECT EXPENSES. Some store expenses, which are pro-rated to all selling departments on the basis of their sales volume, include store maintenance, insurance, and salaries of top management. Additionally, within both the categories of direct and indirect expenses are the CONTROLLABLE and NONCONTROLLABLE expenses.

A. Elements of Each Basic Profit Factor

It is necessary to dissect each of the basic profit factors because each one consists of elements that contribute to profit.

1. Operating Income

a. Gross Sales. Gross sales are the total initial dollars received for merchandise sold during a given period.

CONCEPT:

Gross sales = Total of all the prices charged consumers on individual items × Number of units actually sold.

PROBLEM:

On Monday, an accessories department sold 30 scarves priced at $15 each, 25 scarves priced at $25 each, and 5 scarves priced at $30 each. What were the gross sales for that day?

SOLUTION:

30 pcs. @ $15 each	=	$	450
25 pcs. @ $25 each	=		625
5 pcs. @ $30 each	=		150
Gross Sales	=	$1,225	

b. Customer Returns and Allowances. These are factors because the customer (in either of these transactions) gets a complete refund of the purchase price, or a partial rebate, and a deduction must be made from the gross sales figure because there has been some cancellation of sales. This dollar figure is expressed as a percent of gross sales.

CONCEPT:

Customer returns and allowances = Total of all refunds or credits to the customer on individual items of merchandise × Number of units actually returned.

PROBLEM:

On Saturday, the coat department refunded $98 for one coat, $75 each for two coats, and $55 each for two coats. The other returns for the week amounted to $400; and the weekly total of allowances given was $57. What was the dollar amount of customer returns and allowances?

SOLUTION:

$ 98 × 1	=	$ 98
75 × 2	=	150
55 × 2	=	110
Customer returns for Saturday	=	$358
+		
Total weekly customer returns	=	400
+		
Total weekly customer allowances	=	57
Customer Returns & Allowances (for week)	=	**$815**

CONCEPT:

Customer returns and allowances percentage = The dollar sum of customer returns and allowances expressed as a percent of gross sales.

$$\text{Customer returns \& allowances \%} = \frac{\$\text{Cust. ret. \& allow.}}{\text{Gross sales}}$$

PROBLEM:

Last week the glove department had gross sales of $40,750. The customer returns and allowances for the week totaled $1,630. What was the combined percent of allowances and merchandise returns for the week?

SOLUTION:

$$\text{Customer returns \& allowances \%} = \frac{\$\ 1,630 \text{ Cust. ret. \& allow.}}{\$40,750 \text{ Gross sales}}$$

Customer Returns & Allowances %　　=　　4%

c. Net Sales. NET SALES are the sales total for a given period after customer returns and allowances have been deducted from gross sales.

CONCEPT:

Net sales = Gross sales – Customer returns and allowances.

PROBLEM:

A neckwear department sold $65,000 worth of merchandise. Customer returns were $6,500. What were the net sales of this department?

SOLUTION:

Gross sales	=	$65,000
– Cust. returns & allowances	=	– 6,500
Net Sales	=	**$58,500**

In retailing, the operating income is known as net sales. The net sales figure, also called SALES VOLUME, is used to designate the size of a particular store or a merchandise department. For example, last year, Department #37 had a sales volume of $1,000,000. This dollar figure is generally the sales volume for the year.

Retailers use net sales to measure a department's performance or productivity. It is common practice to calculate the percentage of sales that an individual department has contributed to the total store's net sales. This type of analysis allows a retailer to compare a particular department of one store with other stores within the company, as well as to compare this selected department's sales with industry figures. These figures are published annually by the National Retail Federation in the *Merchandising and Operating Results of Department and Specialty Stores* (MOR).

CONCEPT:

The individual department's net sales are expressed as a percent of the store's total net sales.

$$\text{Department's net sales \% of total store} = \frac{\text{Department's net sales}}{\text{Store's total net sales}}$$

PROBLEM:

The costume jewelry department had net sales of $900,000. For the same period, the total store sales were $45,000,000. What is the percentage of the costume jewelry department's net sales to the total store's net sales?

SOLUTION:

$$\frac{\text{Department's net sales}}{\text{Total store net sales}} = \frac{\$\ \ 900,000}{\$45,000,000}$$

Net Sales % of Total Store $=$ 2%

Because net sales are determined by the adjustment of customer returns and allowances to gross sales, it is also possible through this relationship to calculate, when desired, a gross sales amount if the dollar net sales and the percentage of customer returns and allowances are known.

CONCEPT:

$$\text{Gross Sales} = \frac{\text{Net sales}}{100\% \ (\text{Gross sales}) \ - \ \text{Customer returns \& allowances}}$$

PROBLEM:

The net sales of Department #93 were $460,000. The customer returns and allowances were 8%. What were the gross sales of this department?

SOLUTION:

$$\frac{\text{Net sales}}{100\% - \text{Cust. ret. \& allow. \%}} = \frac{\$460,000}{100\% - 8\%}$$

$$\text{Gross sales} = \frac{\$460,000}{92\%}$$

Gross Sales $= \$500,000$

▶ **For assignments for the previous section, see practice problems 1–10.**

2. Total Cost of Merchandise or Goods (Total Merchandise Costs)

In actual practice, cost of merchandise is affected by the following factors:

▶ Billed cost. Billed cost is the price at which merchandise is purchased and which appears on the invoice (i.e., vendor's bill).

PLUS

▶ Inward freight or transportation charges. This is the amount that a vendor may charge for delivery of merchandise. Inward freight plus billed cost is called the billed delivered cost.

PLUS

▶ Alteration and workroom costs. This is the amount that selling departments may be charged for work that is necessary to put the merchandise in condition for sale, (i.e., assembling, polishing, making cuffs, etc.) It is accepted practice to treat this figure as an addition to cost because it applies only to merchandise that has been sold and not to all purchases.

MINUS

▶ Cash discounts[2]. These are discounts that vendors may grant for payment of an invoice within a specified time. For example, 2% deducted from the total of the billed cost, provided that payment was made within the designated time. The discounts are offered in the form of a percent of the billed cost only, but the dollar discount earned is used in the calculation of the total cost of sales. For example, a 2% cash discount on a billed cost of $1,000 is translated into a $20 deduction:

Billed Cost	$1,000.00
× Cash Discount %	× .02
Dollar Discount	$ 20.00

CONCEPT:

Total cost of merchandise = Billed cost + Inward transportation charges + Workroom costs − Cash discount.

PROBLEM:

An activewear department, for a six-month period, had billed costs of merchandise amounting to $80,000; transportation charges of $2,000; earned cash discounts of 7 1/2%, and had $500 in workroom costs. Find the total cost of merchandise.

SOLUTION:

Billed costs	=	$80,000
+ Inward freight		+ 2,000
Billed Delivered cost	=	$82,000
+ Workroom costs		+ 500
Gross merchandise costs	=	$82,500
− Cash discount (7 1/2% × $80,000)		− 6,000
Total Cost of Merchandise	=	$76,500

▶ **For assignments for the previous section, see practice problems 11–17.**

---

[2]In Unit VI, Invoice Mathematics — Terms of Sale, the other discounts, in addition to cash discounts, that can reduce the cost of goods are discussed.

3. Operating Expenses

Because the expenses of operating a business are deducted from the GROSS MARGIN[3] to determine whether or not a net profit is achieved, the control and management of these expenses are of major concern. For the purpose of analysis the expenses incurred by the retailer (e.g., maintainance of store space, salaries, etc.) are classified to measure the performance of the function or activity. There are various approaches to classifying these items and though there are many different kinds of expenses, each can be easily identified as just that. However, there is a variation in the format used to record them. Traditionally, operating expenses fall into two major categories and are charged to a merchandise department to determine its net profit. These major categories are:

a. Direct Expenses. These are expenses that exist only with a given department and cease if that department is discontinued. These might include: salespeople's and buyer's salaries, buyer's traveling expenses, advertising, selling supplies, delivery to customers, and selling space. For the purpose of expense analysis in retailing, the amount of floor space occupied that generates a given department's sales volume is allotted by the square foot and is charged directly to that department even though there is no cash outlay. Each expense and/or the total direct expenses are expressed as a percent of net sales. For example, if the sales of a department are $100,000 and $3,500 is spent on advertising, the percent of advertising expense would be $3,500 ÷ $100,000 or 3.5%.

b. Indirect Expenses. These are store expenses that will continue to exist even if the particular department is discontinued. These might include store maintenance, insurance, security, depreciation of equipment, and salaries of senior executives. Many indirect expenses are distributed to each department on the basis of its sales volume (e.g., if a department contributes 1.5% to the store's total sales, the indirect expenses charged to this department are 1.5%).

CONCEPT:

Operating expenses = Direct expenses + Indirect expenses

PROBLEM:

A childrens' department has net sales of $300,000 and indirect expenses that are 10% of net sales. Direct expenses are:

- ▶ Selling salaries = $24,000.
- ▶ Advertising expenses = $6,000.
- ▶ Buying salaries = $12,000.
- ▶ Other direct expenses = $18,000.

Find the total operating expenses of the department in dollars and as a percentage.

SOLUTION:

Indirect expenses (10% × $300,000) = $30,000

Direct expenses:

Selling salaries	24,000
Advertising	6,000
Buying salaries	12,000
Other	18,000
Total Operating Expenses	= $90,000

$$\frac{\text{Operating expenses}}{\text{Net sales}} = \frac{\$\,90,000}{\$300,000} = 30\%$$

---

[3]See page 19 for the definition of this term.

c. Controllable and Noncontrollable Expenses. These are expenses that further complicate expense assignments. Many (but not all) direct expenses are controllable. For example, the rent for branch store Y is directly related to this store, but is not under the control of the present store manager because this expense was previously negotiated. Utilities are another example of a direct expense to a store, but indirect to a particular department. The rates are not controllable, but the utilization is.

Because retailers do not always agree on the handling of expenses, some firms use the contribution technique to evaluate the performance of a buyer or store manager. CONTRIBUTION or CONTROLLABLE MARGIN is the difference between gross margin and those expenses that are direct, controllable, or a combination of direct and controllable (e.g., selling salaries). Contribution is the amount that remains after direct expenses are subtracted from the gross margin. It is the amount the department contributes to indirect expenses and profit as seen in Figure 1.

In a learning situation, it is more important to identify expense items, to understand the control, management, and relevancy to profit of those expenses, than to make the accounting decision as to which is direct, indirect, controllable, or noncontrollable. To eliminate the confusion of how to classify and charge a particular expense item, in this text expenses are listed either individually or are referred to as operating expenses.

FIGURE 1 Contribution Operating Statement

		Dollars	Percentages	
Net Sales		$500,000	100.0%	
(minus) **Cost of Merch. Sold**		−266,000	−53.2%	
		$234,000	46.8%	
(minus) **Direct Expenses**				
Payroll	$73,000			
Advertising	13,000			
Supplies	7,000			
Travel	5,000			
Other	12,000			
		−110,000	−22.0%	$\left(\dfrac{\$110,000}{\$500,000}\right)$
Contribution		124,000	24.8%	$\left(\dfrac{\$124,000}{\$500,000}\right)$
Indirect Expenses		106,500	21.3%	$\left(\dfrac{\$106,500}{\$500,000}\right)$
Operating Profit		$17,500	3.5%	

▶ **For assignments for the previous section, see practice problem 18.**

B. The Relationships of the Basic Profit Factors

The following example shows the relationship of the three fundamental factors upon which the amount of profit depends. For comparison, these factors are expressed in percentages as well as in dollars. The net sales figure is the basis for determining profit computation, and so is considered 100%. The other factors involved are shown and stated as a percentage of net sales.

EXAMPLE:

Net sales volume	= $10,000	100%		
(Operating income)				
− Cost of merch. sold	= − 5,500	− 55% →	$\left(\dfrac{\$\ 5,500}{\$10,000}\right)$	
− Operating expenses	= − 4,300	− 43% →	$\left(\dfrac{\$\ 4,300}{\$10,000}\right)$	
Profit	= $ 200	2% →	$\left(\dfrac{\$\ \ \ \ 200}{\$10,000}\right)$	

▶ **For assignments for the previous section, see practice problems 19–24.**

NOTES

1. Customer returns and allowances for department #620 came to $4,500. Gross sales in the department were $90,000. What percentage of merchadise sold was returned?

2. The gross sales for store B were $876,500. The customer returns and allowances were 10%.
    (a.) What was the dollar amount of returns and allowances?
    (b.) What were the net sales?

3. The net sales of department X were $46,780. The customer returns were $2,342. What were the gross sales?

4. The gross sales of store C were $2,500,000. The customer returns and allowances were $11,360. What were the net sales?

5. The net sales of department Y were $36,000. The customer returns and allowances were 10%. What were the gross sales?

6. After Mother's Day this year, the loungewear department had customer returns of 10.5%. The department's net sales amounted to $635,380. As the buyer reviewed last year's figures for the same period, the customer returns were 12.5% with gross sales of $726,149.

   (a.) Mathematically compare the department's performance this year against that of last year, in regard to gross sales, customer returns, and net sales.

   (b.) Discuss the performance from a profit viewpoint.

7. For the year, department store G's total sales amounted to $550,000,000. The junior dress department had net sales of $8,250,000 and the misses dress department had sales of $24,750,000. What were the net sales percentages of each department to the total store?

8. Branch store H had total sales of $30,000,000. The hosiery department sales were 1.9% of store H's total sales. The handbag department's sales were 2.1% of the total branch sales. What were the dollar net sales for each department?

*9. Explain why a merchant should be alarmed if customer returns are excessive. How does a merchant determine what is an excessive percentage of returns? What can be done by the department itself to correct a problem rate of returns?

---

*For research and discussion.

10. The customer returns for the Fall season were 10% on gross sales of $900,000. For the Spring season, gross sales were $850,000 and the customer returns were $70,000. What was the percentage of customer returns for the entire year?

11. A luggage buyer purchased 72 attaché cases that cost $40 each. Cash discounts earned were 2%. Freight charges (paid by the store) were $95. Find the total cost of the merchandise on this order.

12. A misses ready-to-wear buyer placed the following order:

  ▶ 36 pants costing $10.75 each.
  ▶ 48 pants costing $15.50 each.
  ▶ 24 pants costing $17.50 each.

Shipping charges (paid by the store) were 6% of billed cost. Find:

  (a.) The dollar amount of shipping charges.
  (b.) The delivered cost of the total order.

13. A gift shop had workroom costs of $575. The billed cost of merchandise sold amounted to $59,000, with cash discounts earned of $1,180, and freight charges of $650. Find the total cost of the merchandise.

14. A specialty dress shop made purchases amounting to $3,700 at cost, with 8% cash discounts earned, workroom costs of $100, and no transportation charges. Determine the total cost of the merchandise.

15. A sporting goods buyer placed the following order:
    - ▶ 18 nylon backpacks costing $22 each.
    - ▶ 12 two-person tents costing $54 each.
    - ▶ 6 camp stoves costing $55 each.

    Shipping costs, paid by the store, were $60 and a cash discount of 1% was taken. Find:

    (a.) Billed cost on the total order.
    (b.) Total delivered cost of the merchandise.

16. Discussion Problem: Explain why control of inward transportation costs and workroom (alterations) costs is vital. Can a merchandiser help to control these factors? How?

17. Why is cash discount calculated on billed cost?

18. Analyze the expense section of the following statement:

JOHNSTON CANDY COMPANY

Operating Expenses	Last Year	This Year	Plan
Advertising	$ 28,500	$ 24,300	$ 27,700
Sales salaries	74,000	75,100	75,300
Misc. selling costs	8,300	8,300	8,500
TOTAL	$110,800	$ 107,700	$111,500
Net sales	$900,000	$1,140,000	$980,000

(a.) What is the percentage of selling expenses for each year?

(b.) What is the percentage of sales salary expenses for each year?

(c.) What is the yearly percentage of total operating expenses shown?

(d.) What is the trend of operating expenses?

(e.) What is the trend of net sales?

19. Determine the net profit or loss in dollars and percent for a small leather goods department if:

Net sales   = $278,000
Cost of merchandise = $191,600
Operating expenses = $ 78,300

20. Calculate the net profit or loss in dollars and percent if:

Net sales   = $30,000
Cost of merchandise =   52%
Operating expenses =   48%

21. A drug store has net sales of $490,000, with the cost of merchandise sold at $240,000. If the buyer must achieve a 5% net profit, calculate the allowable expenses in dollars and percents.

22. The housewares buyer was allotted an advertising budget of 2.5%. The net sales for this department were planned at $725,000. 60% of the entire advertising dollars was designated for newspaper advertising. What is the dollar amount available for television advertising?

23. The swimwear department of the ABC Department Store had the following records:

Net sales   = $132,000
Cost of merchandise = $118,000
Operating expenses = $ 11,500

What is the net profit or loss?

24. The Rappoport Ribbon Company had the following records:

Gross sales   = $110,000
Customer returns & allowances   = $ 10,000
Cost of merchandise   = $ 80,000
Operating expenses   = $ 35,000

What is the net profit or loss?

► II. PROFIT AND LOSS STATEMENTS

Businesses must keep an accurate record of sales income, merchandise costs, and operating expenses to calculate profit. Periodically, the income and expenses are summarized on a form known in retailing as a STATEMENT OF PROFIT AND LOSS. (In other types of organizations, this statement is frequently called an INCOME STATEMENT.) It is a summary of the business transactions during a given period of time, in terms of making or losing money. Generally, the accounting department of the business keeps a continuous record of the sales income and the expenses, and at set intervals a statement is compiled that shows whether these transactions result in a profit or a loss. (The interval it covers might be a year, three months, or one month.) A profit and loss statement should not be confused with a balance sheet that shows the assets, liabilities, and net worth of a business.

Because this is a retailing text, the profit and loss statement will not be analyzed as a bookkeeping procedure, but in terms of how the data it contains can be used by a merchant to improve a merchandising operation. It is a fundamental merchandising concept that one of a buyer's chief responsibilities is to ensure that a store or department earns a profit. It is essential to remember that a profit is earned only on the merchandise that is sold during the accounting period under consideration.

A profit and loss statement shows the difference between income and expense(s). If income exceeds expenses, the result is a profit. If expenses exceed income, the result is a loss. The five major components of a profit and loss statement are:

- ► Net sales.
- ► Cost of merchandise sold.
- ► Gross margin (the difference between net sales and cost).
- ► Operating expenses.
- ► Profit or loss figure.

The difference between net sales and the total cost of merchandise sold is the GROSS MARGIN. This figure must be large enough to cover operating expenses and to allow for a reasonable profit. It is calculated for a given period by subtracting the total cost of the merchandise sold from the net sales for the period. Gross margin is also frequently called GROSS PROFIT because it is an indicator of the final results.

CONCEPT:

Gross margin = Net sales – Total cost of merchandise sold.

PROBLEM:

A department had net sales of $300,000 with the total cost of merchandise sold at $180,000. Find the dollar gross margin.

SOLUTION:

Net sales	=	$300,000
– Cost of merchandise sold	=	–180,000
Gross Margin	=	$120,000

A. Skeletal Profit and Loss Statements

A SKELETAL PROFIT AND LOSS STATEMENT does not spell out in detail all of the transactions, but is a quick method to determine, at any particular time, any given department's profit and loss. It only contains the five major components and is expressed in both dollars and percents.

EXAMPLE:

Net sales	=	$10,000	100%
– Cost of merchandise sold	=	– 6,000	– 60%
Gross margin	=	$ 4,000	40%
– Operating expenses	=	– 3,500	– 35%
Net Profit	=	$ 500	5%

B. The Skeletal Profit and Loss Statement Expressed in Percentages

The value of a profit and loss statement is that it can be used as a comparison with previous statements or against industry-wide figures to help improve profit, or any of the other factors mentioned in the example. Therefore, it is vital to think in terms of percentages in addition to the dollar amounts. For example, if a buyer were to declare a net profit of $2,869 for a given business period, this statement would have no real meaning unless the dollar amount for every one of the other contributing factors was also stated. The profit figure could be phenomenally high or dismally low by industry standards, depending on the dollar net sales volume of the department. Unless all the the other factors are available, the determination of which departmental operations had excelled or faltered would be impossible. The only meaningful way to compare departmental performances is to compare the respective results expressed as a percentage of the net sales volume. From this, the deduction can be made that profit will vary upward or downward as one or more of the three major factors vary (i.e., net sales, cost of merchandise sold, or operating expenses).

CONCEPT:

$$\text{Cost of merchandise sold \%} = \frac{\text{Cost of merchandise sold (in dollars)}}{\text{Net sales}}$$

$$\text{Gross margin \%} = \frac{\text{Gross margin (in dollars)}}{\text{Net sales}}$$

$$\text{Operating expenses \%} = \frac{\text{Direct \& indirect expenses (in dollars)}}{\text{Net sales}}$$

$$\text{Net profit \%} = \frac{\text{Net profit (in dollars)}}{\text{Net Sales}}$$

PROBLEM:
The junior sportswear department in Store A had net sales of $160,000; cost of goods sold was $88,000; operating expenses were $64,000. The junior sportswear department in store B, for the same business period, had net sales of $260,000; cost of goods sold was $135,200; operating expenses were $109,200. Which store produced a higher net profit percentage?

SOLUTION:

	STORE A		STORE B	
Net sales	$160,000	100%	$260,000	100%
− Cost of goods sold	− 88,000	− 55%	−135,200	− 52%
Gross margin	$ 72,000	45%	$124,800	48%
− Oper. expenses	− 64,000	− 40%	−109,200	− 42%
Net Profit	$ 8,000	5%	$ 15,600	6%

As a basis for comparison, the percentage figures give the clearest picture. Upon examination of this skeletal profit and loss statement, the reader can see that the junior sportswear department in store A spent 55 cents of every dollar of sales on the cost of merchandise sold, and store B spent 52 cents. Respectively, store A spent 40 cents and store B spent 42 cents of every dollar of sales on operating expenses. Gross margin, net profit, and individual transactions are more easily comparable when these figures are recorded in a complete profit and loss statement. Also, all other figures in the skeletal profit and loss statement are expressed as a part or a percentage of net sales.

▶ For assignments for the previous section, see practice problems 25–34.

C. Final Profit and Loss Statements

For a fully developed profit and loss statement, first the basic factors (i.e., sales, cost of goods, expenses) were analyzed. Then a skeletal statement that is the framework of an amplified statement, providing a quick method of monitoring profit or loss, was devised. Figure 2 illustrates a FINAL PROFIT AND LOSS STATEMENT, showing the basic profit factors developed in detail so that every transaction is clearly seen. It is necessary that the many factors be in a standard arrangement so that it is possible to have an easily analyzed picture of the results.

A final profit and loss statement includes the opening and closing figures of STOCK-ON-HAND because the firm only realizes a profit on merchandise that is sold during a particular accounting period. This is the only way that this statement can be used as a comparison with other stores, the only way one can detect weaknesses that need strengthening, and it is an excellent measure for illuminating those strengths which bear repetition.

To determine the cost of only the merchandise that was sold, the final profit calculation requires a TOTAL MERCHANDISE HANDLED amount, which is the sum of the merchandise on hand, or OPENING INVENTORY plus the new purchases. It can be determined at cost or at retail. The total cost of the new purchases uses the invoice or billed costs of these purchases to which is added transportation charges. The closing inventory, or merchandise remaining at the end of the accounting period is then deducted. For example:

Opening inventory, at cost	$100,000
+ Billed costs	+500,000
+ Inward freight	+ 1,000
Total merchandise handled, at cost	$601,000
− Closing inventory, at cost	−150,000
Gross Cost of Merchandise Sold	$451,000

Now it is possible to find the NET COST OF MERCHANDISE SOLD by the other adjustments (e.g., cash discounts). For a better understanding of the detailed Profit and Loss Statement shown on page 22, refer to the glossary of terms that accompanies Figure 2.

FIGURE 2 Profit and Loss Statement

Profit Factors	Cost		Retail	%
Income from sales				
Gross Sales			$450,000	
– Customer Returns & Allowances			–25,000	
Net Sales			$425,000	(100%)
Cost of Merchandise Sold				
Opening Inventory		$ 52,000	$100,000	
New Net Purchases	$ 258,000			
+ Inward Transportation	+ 2,000			
	$260,000			
Total Cost of Merchandise		$260,000		
Total Merchandise Handled		$312,000		
– Closing Inventory		– 65,000		
Gross Cost Mdse. Sold		$247,000		
– Cash Discount		– 13,000		
Net Cost of Mdse. Sold		$234,000		
+ Alteration & Workroom Costs		+ 1,000		
Total Merchandise Cost		$235,000	–235,000	(55.3%)
GROSS MARGIN			$190,000	(44.7%)
Operating Expenses				
Total Direct Expenses		$101,250		
Total Indirect Expenses		67,500		
Total Operating Expenses			$168,750	(39.7%)
Net Profit			$ 21,250	(5%)

GLOSSARY OF TERMS

▶ **INCOME FROM SALES** is divided into such categories as:

Gross Sales ($450,000), which are the retail values of total initial sales.

Customer Returns and Allowances ($25,000), which represent cancellation of sales by either customer credit, refund, or partial rebate.

Net Sales Figure ($425,000), which is derived by subtracting customer returns and allowances from the gross sales of a period. It is the dollar value of sales that "stay sold."

▶ **COST OF MERCHANDISE SOLD** is divided into such categories as:

Opening Inventory, At Retail ($100,000), which is the amount of merchandise at the beginning of a period, counted and recorded at the current selling price.

Opening Inventory, At Cost ($52,000), which is derived from the retail by applying a markup percentage on the total merchandise handled.

New Net Purchases ($258,000), which represents the billed cost of merchandise purchased. The gross purchases minus returns and allowances to vendors.

Inward Transportation ($2,000), which is the cost of transporting the goods to the premises.

Total Cost of Merchandise ($260,000), which combines the cost of merchandise purchased and inward transportation.

Total Merchandise Handled at Cost ($312,000), which is the sum of the opening inventory plus the total cost of the purchases.

Closing Inventory, at Cost ($65,000), which is derived from the retail inventory figure and represents the merchandise in stock at the end of an operating period.

Gross Cost of Merchandise Sold ($247,000), which represents the total merchandise handled less the cost of the closing inventory.

Cash Discounts ($13,000), which are adjustments made to the cost of goods sold from paying bills in a specified time.

Net Cost of Merchandise Sold ($234,000), which is the gross cost of merchandise sold minus cash discounts.

Alteration and Workroom Costs ($1,000), which is another adjustment to the cost of goods sold and refers to the cost of preparing goods for resale.

Total Merchandise Costs ($235,000 or 55.3%), which results from subtracting cash discounts and adding alteration and workroom costs to gross cost of merchandise sold.

Gross Margin ($190,000 or 44.7%) is the difference between the net sales and the total merchandise cost. This is often called gross profit.

▶ **OPERATING EXPENSES** are divided into such categories as:

Direct Expenses ($101,250), which come into being with a department and cease if it is discontinued.

Indirect Expenses ($67,500), which will continue to exist even if a department is discontinued.

Total Operating Expenses ($168,750), which are direct and indirect expenses combined.

▶ **NET (OPERATING) PROFIT** ($21,250 or 5%) are the results from the relationship of sales, cost of goods and expenses. When the gross margin is larger than the operating expenses, a net profit is achieved.

NOTES

25. Note the following figures:

    Net profit   2.5%
    Gross margin $7,000
    Operating expenses $6,600

    Find:

    (a.) The cost of goods in dollars.

    (b.) The percentage of operating expenses.

26. The net profit in an appliance department for the Spring/Summer period was $20,000, which represented 2% of net sales. Operating expenses totaled $480,000. Find:

    (a.) The dollar amount of gross margin.

    (b.) The net sales figure.

27. The linen department had net sales of $80,000. There was a 2% loss and the gross margin was 46.5%. What were the operating expenses of the department in dollars and in percents?

28. Calculate the percentage of operating expenses for a home furnishings department that has the following figures:

Gross sales	$476,000
Customer returns	4,000
Advertising costs	10,000
Salaries	101,000
Miscellaneous expenses	6,160
Utilities	9,000
Insurance	11,000
Rental	70,000

29. Set up a skeletal statement that shows both dollars and percentages for the following figures:

Net profit	$5,500
Net profit	2.5%
Operating expenses	47.5%

30. Suppose that the estimated net sales for the coming year are $100,000; estimated cost of merchandise purchases is $52,000; the total estimated operating expenses are $43,000. The buyer wants a net profit of 5%. Determine the percentage of gross margin on sales needed to achieve this desired profit.

31. Using the following figures, set up a skeletal profit and loss statement that shows each factor in dollars and percents.

Net sales	$85,000
Net profit	1,700
Cost of goods sold	45,000

32. What is the profit or loss in dollars if gross sales are $218,000, customer returns and allowances are $3,000, cost of goods sold come to 55%, and the operating expenses are 41%?

33. Find the gross margin percent when:

Gross sales	$435,000
Customer returns and allowances	49,000
Billed cost of goods	195,000
Freight charges	1,800
Cash discounts	4%

34. Discussion Problem: In measuring gross margin performance, which is more significant: the dollar amount or the percentage figure? Why?

FIGURE 3 Profit Performance by Store Type

Store Type	Profits		Profits to Sales		
	Last Year	This Year	% Change	Last Yr	This Yr
Department stores	$1,470,655	$1,580,722	+ 7.5	3.3%	3.5%
Mass merchandisers	1,896,700	2,033,300	+ 7.1	2.8	2.8
Specialty stores	555,579	568,774	+ 2.4	4.4	5.0
Discount stores	335,336	432,774	+29.1	3.4	3.3
Off-price stores	24,143	32,073	+32.9	5.4	5.2
Miscellaneous	478,326	461,537	− 3.5	2.3	2.6
Total	$4,762,739	$5,109,000	+ 7.3	3.0	3.2

III. HOW TO INCREASE PROFITS

Realistically, it is impossible to list everything a retailer needs to know to merchandise at a profit because the factors that govern profits are variable and net profits do not represent any fixed sum. As a frame of reference, Figure 3 lists various types of retail stores and their profit performance. These results illustrate that the amount of profit generated can be different depending on the type of organization and is rarely, if ever, constant.

After reviewing a profit and loss statement, however, and seeing how these factors affect profit, certain measures can be taken to improve profits. Because there is always an interrelationship between the three factors (i.e., sales, cost of merchandise sold, and operating expenses), the adjustments made must balance all three factors in relationship to each other. Fundamentally, an approach to improved profits can be to:

- ▶ Increase sales while there is only a proportionate increase in the cost of the merchandise, and little or no increase in expenses.
- ▶ Decrease the cost of merchandise sold without a decrease in sales (e.g., selling a larger proportion of higher markup merchandise, or decreasing the net cost of goods sold by lowering shipping charges and/or obtaining greater cash discounts).
- ▶ Reduce expenses.

The following example shows the application and effect of each approach:

For the accounting period under consideration, a merchant estimated sales for $100,000, estimated merchandise purchases at $70,000 and estimated total operating expenses at $25,000. If the merchant wants to obtain more than the previous 5% net profit, what approach can be taken?

ACTUAL ESTIMATED PERFORMANCE

	Dollars	Percentages
Sales	$100,000	100%
− Cost of merchandise sold	− 70,000	− 70
Gross margin	30,000	30
− Operating expenses	− 25,000	− 25
Net profit	$ 5,000	5%

APPROACH #1:

Increase sales with only a proportionate increase in cost of merchandise sold and little or no increase in expenses.

	Dollars	Percentages	
Sales	$110,000	100%	(Increased sales)
− Cost of mdse. sold	− 75,900	− 69	(Decreased % of cost of mdse. sold)
Gross margin	34,100	31	(Increased $ and % of gross margin)
− Operating expenses	− 28,050	− 25.5	(Increased $ and % of oper. exp.)
Net profit	$ 6,050	5.5%	(Increased net profit)

APPROACH #2:

Decrease cost of merchandise sold without decreasing sales, which is equivalent to a larger gross margin.

	Dollars	Percentages	
Sales	$100,000	100%	(Constant)
− Cost of mdse. sold	− 69,500	− 69.5	(Decreased cost of mdse. sold)
Gross margin	30,500	30.5	(Larger gross margin)
− Operating expenses	− 25,000	− 25	(Same expenses)
Net profit	$ 5,500	5.5%	(Increased net profit)

APPROACH #3:

Lower or reduce expenses.

	Dollars	Percentages	
Sales	$100,000	100%	(Constant)
− Cost of mdse. sold	− 70,000	− 70	(Constant)
Gross margin	30,000	30	(Constant)
− Operating expenses	− 24,500	− 24.5	(Reduced expenses)
Net profit	$ 5,500	5.5%	(Increased net profit)

PRACTICE PROBLEMS

35. Review this store's performance by analyzing the following figures:

Opening inventory, at cost	$120,000
New purchases (billed cost)	140,000
Transportation charges	1,000
Closing inventory, at cost	122,000

   (a.) Determine the total merchandise handled at cost.

   (b.) What is the gross cost of merchandise sold?

36. A retailer, contemplating a purchase of a small children's shop, found the previous owner had an inventory at cost of $30,000. For that period, this retailer was given the following figures:

Sales	$ 60,000
Closing inventory, at cost	33,000
New purchases, at cost	40,000
Transportation charges	500

What was the gross cost of merchandise sold?

37. During the year, a stationery store achieved the following results:

Sales	$585,000
Purchases, at cost	548,000
Transportation charges	5,000

Inventories at cost:

Beginning of year	555,000
End of year	500,000

Determine the gross cost of merchandise sold.

38. Find the net profit or loss as a percentage, and the gross margin as a dollar amount:

Gross sales	$200,000
Customer returns & allowances	15,000
Opening inventory (at cost)	38,000
Billed cost of goods	99,000
Inward transportation	5,000
Cash discount	6,000
Closing inventory (at cost)	36,000
Payroll	48,000
Occupancy	28,000
Wrapping and packing	1,200
Utilities	2,000
Delivery	2,800

39. Construct a profit and loss statement using the following departmental figures and showing the net sales, total cost of goods sold, gross margin, expenses and profit:

Gross sales	$82,000
Customer returns & allowances	4,000
Inward freight	2,000
Workroom costs	1,000
Opening inventory (at cost)	17,000
Closing inventory (at cost)	14,000
Purchases (at cost)	36,000
Cash discounts	8%
Advertising	5,000
Rent	12,000
Salaries	17,000
Miscellaneous expenses	2,500

40. Construct a final profit and loss statement from the figures listed below and calculate the major factors as percentages and dollar amounts.

Opening inventory	$ 74,200
Gross sales	248,000
Advertising	15,000
Misc. expenses	18,000
Purchases (at cost)	120,000
Closing inventory	78,000
Customer returns	25,800
Salaries	26,000
Transportation charges	8,000
Rent	39,000
Cash discounts	3%

41. The following figures are for your department:

Net sales	$490,000
Billed cost of goods	265,000
Freight charges	11,160
Rent	56,640
Salaries	111,600
Miscellaneous expenses	25,260
Cash discounts	37,000
Insurance	27,800
Advertising	16,740
Opening inventory (cost)	117,000
Closing inventory (cost)	120,000

   (a.) Calculate the total cost of merchandise sold.

   (b.) Find the operating expenses in dollars.

   (c.) Determine the profit or loss in dollars and as a percentage.

42. If profit is $8,000 and profit percentage is 4%, what is the net sales figure?

43. If expenses are $85,340 and the gross margin is $90,960, what is the operating profit or loss in dollars?

44. Find the net profit or loss in dollars and as a percentage if:

Opening inventory (cost)	$ 70,000
Operating expenses	180,000
Closing inventory	72,000
Net sales	400,000
Inward freight	5,000
Purchases, at cost	210,000

45. Gross sales are $25,619; customer returns and allowances are $2,791.32; purchases at billed cost are $12,585; inward freight is $932.45; cash discounts earned average 2%. What is the gross margin in dollars?

46. If a department experiences a loss of 3% for a six-month period and its gross margin is 43%, what must be its expense percentage?

47. Gross margin in the shoe department was $185,000. Operating expenses were $178,000 and the net profit was 2% of net sales. What were the net sales?

48. Determine the net sales when:

Operating expenses	$57,750
Gross margin	$56,650
Net loss	1%

49. Find the gross margin as a percentage if:

Gross sales	$283,000
Customer returns	7,000
Billed cost of goods	137,000
Inward freight	5,000
Workroom charges	3,000
Cash discounts	6%

50. Set up a final profit and loss statement using the following figures, giving both dollar amounts and percentages for:

(a.) Net profit or loss.

(b.) Cost of goods sold.

(c.) Gross margin.

(d.) Operating expenses.

Inward freight	$ 3,000
Workroom and alteration charges	600
Opening inventory (at cost)	14,400
Closing inventory (at cost)	14,600
Customer returns	8,000
Gross sales	82,000
Purchases (at cost)	35,000
Promotional expense	3,500
Rent and utilities	10,000
Payroll	15,500
Miscellaneous expenses	2,000
Cash discounts	3%

Analyze the above figures. Discuss the relationship of expenses to net profit; discuss the size of the gross margin. Suggest ways to improve profit.

51. Determine the dollar profit or loss in a department in which sales were $1,950,000, cost of merchandise sold was 50%, and operating expenses were $925,000. ·

52. Prepare a skeletal profit and loss statement showing both dollars and percents for a department showing the following figures:

Net sales	$280,000
Cost of merchandise sold	150,000
Loss	3%

53. The handbag department shows the following figures:

Gross margin	$11,500
Operating expenses	11,000
Net profit	2%

Find the net sales.

54. A boutique shows these figures:

Gross sales	$104,000
Customer returns	5,000
Billed costs	49,000
Inward freight	10,000
Advertising	7,500
Rent	12,000
Alterations	1,000
Salaries	16,000
Misc. expenses	5,000

Prepare a profit and loss statement. Did this store operate at a profit?

55. Find the gross margin percentage when:

Gross sales	$566,000
Customer returns	10%
Billed cost of goods	258,000
Inward freight	9,000
Workroom charges	2,200
Cash discounts	6%

56. Set up a profit and loss statement, showing dollar amounts only, using the following figures:

Gross sales	$127,000
Closing inventory (at cost)	29,000
Opening inventory (at cost)	33,000
Miscellaneous expenses	5,000
Customer returns	6,000
Purchases (at cost)	60,000
Rent and utilities	23,000
Payroll	30,000
Transportation charges	1,700
Cash discounts	8%
Advertising	6,000

57. Construct a skeletal profit and loss statement from the figures below, expressing all factors in both dollars and percentages.

Billed cost of goods	$ 90,000
Alteration & workshop charges	2,400
Cash discounts	1,800
Freight charges	3,000
Direct expenses	36,000
Customer returns	9,000
Indirect expenses	31,200
Gross sales	177,000

58. Determine the percent of profit or loss if:

Gross sales	$180,000
Direct expenses	52,000
Opening inventory (at cost)	39,000
Indirect expenses	27,000
Purchases, at cost	95,000
Customer returns	8,000
Inward freight	1,000
Closing inventory (at cost)	44,000
Cash discounts	4%

*59. Your merchandise manager has pointed out that the profit percentage in your department for the most recent period was extremely poor. In conjunction with your assistant buyers, make a list of as many ways that you, as the buyer, can attempt to improve the situation directly.

---

*For research and discussion.

*60. Set up a mathematical demonstration example to show how increased profit percentage might be achieved in a departmental operation despite reduced volume. Explain.

61. Based on what you already know as a consumer, and in view of changing economic patterns over the last three decades, make a "ball park" estimate of the cost, gross margin, and expense figures in the spaces below. Give a brief explanation of your estimated figures.

	1960-1970	1970-1980	1980-1990
Net sales	$1,000,000	$1,000,000	$1,000,000
– Cost of goods	_____	_____	_____
= Gross margin	_____	_____	_____
– Expenses	_____	_____	_____
= Profit	$ 30,000	$ 30,000	$ 30,000

*For research and discussion

Merchandising for a Profit

For the past two years, Miss Rose has been the owner of "My Secret Love", a trendy boutique specializing in moderate-priced handbags. Previously an assistant buyer in a major department store, her training and experience were instrumental in the development of a successful and profitable business. Lately, however, she realized that the current economic conditions were less favorable. Sales were sluggish and she could see that maintaining her expenses would become increasingly difficult. She decided for the next season to target another customer group and add a high-priced designer line to her assortment. To properly merchandise this new classification, she wants to hire another part-time salesperson skilled in selling this type of goods, and to use additional sales promotion techniques to launch this group. Most important, she wants to retain the profit percent she has achieved, though generally sales are soft and her expenses will increase.

In the first year, "My Secret Love's" net sales were $450,000 with a 4.5% profit. Miss Rose was pleased with her venture. The second year her sales grew to $481,500 and her profit improved to 5.0% with expenses totaling $192,600 as opposed to 40.5% her first year. She felt she was moving in the right direction. Now, as she anticipates the future, she hopes to increase her sales by 7%, and she estimates her expenses to reach 41%.

(a.) How, if at all, will her change in sales and expenses affect her profit?

(b.) As you review Miss Rose's past and future goals, what suggestions can you offer so that she can achieve them?

Retail Pricing and Repricing of Merchandise

- ► Identification of activities that retailers can use to maximize profits.
- ► Understanding of price lining and identification of the types of price zones.
- ► Recognition and identification of the three basic pricing elements, and description of their relationship.
- ► Calculation of markup as dollar and percentage amounts for individual items and groups of items.
- ► Establishment of retail prices.
- ► Ascertainment of the types of price adjustments and confirmation of their importance as merchandising decisions.
- ► Calculation of markdowns as dollar and percentage amounts.
- ► Delineation of the procedures for making price changes.
- ► Recognition of the impact of pricing and repricing decisions on profit.

KEY TERMS

additional markup

aggregate sales

cost complement

employee discount

gross markdown

markdown

markdown cancellation

markdown percentage

markup

markup cancellation

markup percentage

net markdown

point-of-sale markdown

prestige price zone

price lining

price range

price zone

promotional price zone

retail reductions

volume price zone

One of the aims of every business is to yield the largest possible total profit. One of the ways a retail merchandiser attempts to secure maximum profits is through the skillful pricing of goods offered for sale. Price is a strong motivation in consumer buying habits. It is a competitive weapon and very frequently it is the only way to attract customer patronage when merchandise assortments are comparable, if not identical. Because there are many factors that influence pricing, it can be considered an art as well as a science.

In large industrial organizations, the actual pricing decisions of products are generally the responsibility of management. In large retail stores, the actual pricing of merchandise offered for sale is determined by the individual departmental buyers or another comparable person designated by the particular organizational structure. Top management, however, does formulate the basic price policies of the store; for example, implementing a policy of underselling the competition by 10% on all items. Although the retailer establishes the price of individual items as they are offered for sale, ultimately, the total of all purchases must realize maximum profits. It must be remembered that in the final analysis, the volume of sales as an aggregate figure must be great enough to cover not only the costs of merchandise sold, but also must provide the reward of profit. Pricing, therefore, is an integral part of merchandising that requires training and skill.

▶ I. RETAIL PRICING

Pricing refers to PRICE LINING, which is the practice of predetermining the retail prices at which an assortment of merchandise will be carried. A retail buyer selects and offers a merchandise assortment to the consumer at a specific price point, or PRICE LINE; for example, $15, $20, etc.

A. The Structuring of Price Lines

A buyer creates a stock assortment by considering what price lines to carry and the depth of assortment offered at the various price points. The number of price lines, and those particular price lines in the assortments, can help reflect the desired character that management wishes to project. The emphasis of a stock by price lines depends on the composition of the consumer segment that management wishes to attract. Usually, there is a price structure around which buying for a specific retail establishment is concentrated. For example, the sportswear department stocks tops to retail for $25; this price line may cover a variety of types, fabrics, and sizes. The buyer, with the proper sources of information, knows that particular store's customers will pay $25 for a top.

The sportswear department, however, may stock a variety of tops that retail from $8 to $40. This is called the PRICE RANGE, which refers to the spread from the lowest to the highest price line carried. Most customers generally prefer to concentrate their purchases at either one price line or several that are relatively close to each other. PRICE ZONE refers to a series of price lines that are likely to appeal to one group of the store's customers. When more than two price lines are stocked, a price zone situation exists. The price zones can be referred to as VOLUME PRICE ZONE, PROMOTIONAL PRICE ZONE, and PRESTIGE PRICE ZONE. For example, when the price range is from $8 to $40, the three price zones can be illustrated in figure 4:

FIGURE 4 Price Line Structure Chart

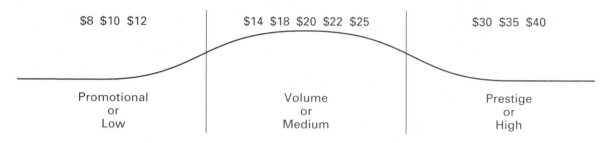

$8 $10 $12	$14 $18 $20 $22 $25	$30 $35 $40
Promotional or Low	Volume or Medium	Prestige or High

Generally, the promotional price zone refers to the lower price lines carried, with the volume price zone in the middle price lines where the largest percentage of sales occur. The prestige price zone refers to the highest price lines carried in a department, which "tones up" the assortment. This distribution curve typically occurs within a price range, regardless of the price lines.

B. Setting Individual Retail Prices and/or Price Lines

The pricing of individual items and the establishment of price lines require experience and skill. When pricing merchandise as it is bought, a buyer must always regard its salability in terms of its retail price, remembering that the aggregate of all prices must be high enough to cover merchandise costs, operating expenses, a fair profit return, possible reductions, shortages, and discounts to employees. Variations in pricing occur because there are some noncontrollable factors that influence pricing decisions, such as the volatility of ready-to-wear. Because they influence the setting of the retail price, the basic factors to consider in pricing are:

- ▶ Wholesale costs.
- ▶ Competition.
- ▶ Price maintenance policies of manufacturers, for example, "suggested" retail prices.
- ▶ Handling and selling costs.
- ▶ Store policies, for example, "off-price" policy.
- ▶ Nature of the goods, for example, markdown risk in fashion goods.
- ▶ Correlation among departments.
- ▶ Supply and demand factors.

C. Advantages of Price Lining

The practice of offering merchandise for sale at a limited number of predetermined price points creates several merchandising problems, but the practice is prevalent because the advantages are numerous and significant. The advantages of price lining are that it:

- ▶ Simplifies customer choice, which facilitates selling.
- ▶ Enables the store to offer wide assortments at best-selling price lines.
- ▶ Simplifies buying by limiting the range of wholesale costs.
- ▶ Reduces the size of stock, resulting in more favorable stock turnover and decreased markdowns.
- ▶ Simplifies stock control.
- ▶ Decreases marking costs.
- ▶ Provides a basis for stock/sales control of customer preferences by price.

D. Basic Pricing Factors and Their Relationship

Merchandising, or the act of buying and selling, is performed by the retailer, who selects and buys merchandise that is offered for resale to the consumer. To make a profit, the retail merchandiser must set the proper price on the merchandise. Though the profitable pricing of selected individual items cannot always be done by applying a mathematical formula, there are three basic elements involved in the pricing of all goods. These elements are the cost of the merchandise, the retail price, and the difference between them, which is referred to as MARKUP (MU). Markup is the amount that is added to the cost price of merchandise to arrive at a retail price. This amount must be large enough to cover the cost of the merchandise, expenses incurred to sell it, and the desired profit. The aim of proper pricing can be expressed by the following illustration:

$$\left.\begin{array}{l}\text{Cost of goods} \\ +\ \text{Markup} \\ \hline = \textbf{Retail Price}\end{array}\right\} \left\{\begin{array}{l}\text{Expenses} \\ \text{Markdowns, shortages, employee discounts}^1 \\ \text{Profit}\end{array}\right.$$

Given any two of the basic pricing factors, the third can be calculated both in dollar amounts and/or in percentage relationships. The following formulas show the relationship, in dollars, of the three basic pricing elements.

1. Calculating Retail When Cost and Dollar Markup Are Known

CONCEPT:
  Retail = Cost + Dollar markup

PROBLEM:
  A retailer buys a top for $10 and has a markup of $8. What is the retail price?

SOLUTION:

Cost	=	$10
+ Markup	=	+ 8
Retail Price	=	$18

2. Calculating Dollar Markup When Retail and Cost Are Known

CONCEPT:
  Dollar markup = Retail − Cost

PROBLEM:
  A retailer buys a top for $10 and decides to price it for $18. What is the dollar markup on this item?

SOLUTION:

Retail	=	$18
− Cost	=	−10
Dollar Markup	=	$ 8

---

[1] Markdowns, shortages, and employee discounts must be considered in determining the original retail price set when merchandise is received into stock because these factors cause a reduction in the value of total purchases.

3. Calculating Cost When Retail and Dollar Markup Are Known

CONCEPT:

Cost = Retail – Dollar markup

PROBLEM:

What is the cost of an item that retails for $18 and has a dollar markup of $8?

SOLUTION:

Retail	=	$18
– Dollar Markup	=	– 8
Cost	=	$10

Although retailers pay for and sell merchandise in dollars, knowing the relationship of the three basic pricing elements (i.e., retail, cost, markup) in percentages and dollars is useful and meaningful to the merchandiser's function. Frequently, the retailer thinks in terms of dollars only, for example, "My customers prefer tops that retail for $18 to $25", or a resource quotes the cost of an item to a retailer at $10 each. At other times, the retailer is concerned about achieving a 44.4% markup on an item that retails for $18 and costs $10. If this markup percent is perceived as low, it may determine whether or not a particular item will be purchased. At still another time, the buyer thinks simultaneously in terms of dollars and percentages. For example, in determining the dollar cost of an article to be purchased, a buyer must know the retail price the customer is willing to pay, and balance this knowledge with a particular markup percent required by management.

4. Calculating Cost and Markup Percentages when Dollar Cost and Retail are Known

To fully understand the relationship of the three basic elements in percentages, it should be understood that the retail price of the basic equation is always 100%, and the dollar cost and markup can be converted from dollars to percentages by expressing each as a part of the retail.

CONCEPT:

$$\frac{\$ \text{ Cost}}{\$ \text{ Retail}} = \text{Cost } \%$$

$$\frac{\$ \text{ Markup}}{\$ \text{ Retail}} = \text{Markup } \%$$

PROBLEM:

What is the cost percent, and the markup percent of a top that costs $10 and retails for $18?

SOLUTION:

Cost %   = $\dfrac{\$10 \text{ Cost}}{\$18 \text{ Retail}}$

Cost %   = 55.6%

Markup %   = $\dfrac{\$8 \text{ Markup } (\$18R. - \$10C.)}{\$18 \text{ Retail}}$

Markup %   = 44.4%

Therefore:

	Dollars	Percents	
Retail	$18	= 100.0%	
– Cost	–10	= 55.6% →	$\left(\dfrac{\$10}{\$18}\right)$
Markup =	$ 8	= 44.4% →	$\left(\dfrac{\$ 8}{\$18}\right)$

The markup percent is commonly called the COST COMPLEMENT.

1. A merchant buys a sweatshirt for $15 and has a dollar markup of $14.95. What is the retail price?

2. The infants' department buyer purchases some wool pram robes that cost $30 each and prices them for $65 each. What is the dollar markup on this item?

3. What is the cost of a set of golf clubs that retails for $250 and has a markup of $157?

4. What is the markup percent on a coat that costs $49.75 and retails for $125?

5. What is the cost percent for a sweater that retails for $38 and has a
   55% markup?

6. Fill in the blank spaces for each of the exercises below.
   (Note: Several of the examples may be done without written calculations.)

	Retail	Cost	$MU	MU%
a.	$5.00	$2.70		
b.	$14.95	$7.50		
c.	$200.00		$90.00	
d.	$1.75	$9.60/doz.		
e.	$29.95	$14.50		
f.	$10.00	$60.00/doz.		
g.		$15.00 each		48%
h.		$42.50 each		46.6%
i.		$106.00/doz.		45%
j.		$15.00/doz		47.5%
k.	$100.00			49%
l.	$80.00			50%
m.		$16.00	$14.00	
n.		$55.00	$45.00	
o.	$75.00			52%
p.	$3.98			37.8%
q.		$21.60/doz.		46.5%
r.	$9.95	$72.00/doz.		
s.	$465.00	$235.00		
t.	$590.00	$310.00	$280.00	

► II. MARKUP

Markup, as defined in this text, applies to an individual item, a group of items, or the entire merchandise stock of a department or a store, and can be expressed in either dollars or percents. In practice, it is the markup percentage that is significant (rather than the dollar amount) for comparison and analysis. Knowing the basic markup equations and their relationship to the other pricing factors aids in the understanding of the effect of markup on buying decisions.

A. Basic Markup Equations

Markup percentages can be figured as a percentage of the retail price or the cost price. Because the retail method of inventory is prevalent in large stores, the markup percentage calculated on the retail price is more common. The cost method of calculating markup is considered to be relatively old-fashioned, although some retailers still use this method.

1. Calculating Markup Percentage on Retail Using the Retail Method of Inventory

CONCEPT:

$$\text{Markup \%} = \frac{\text{Dollar markup}}{\text{Retail}}$$

PROBLEM:

What is the markup percent on an item when the markup is $8 and the retail is $18?

SOLUTION:

$$\text{MU \%} = \frac{\$ 8 \text{ MU}}{\$18 \text{ R.}}$$

$$= 44.4\%$$

2. Calculating Markup Percentage on Cost

CONCEPT:

$$\text{Markup \%} = \frac{\text{Dollar markup}}{\text{Cost}}$$

PROBLEM:

What is the markup percentage on cost, when the markup is $8 and the cost is $10?

$$\text{MU \%} = \frac{\$ 8 \text{ MU}}{\$10 \text{ C.}}$$

SOLUTION:

$$\text{MU \%} = 80\%$$

(Note: Markup percentage calculated on cost is higher than markup percentage on retail.)

Generally, the retail calculation of markup is more acceptable because all expenses and profits are also figured as a percentage of retail sales. Retailers tend to channel their planning of price lines, stocks, and customer demands in retail values, so the calculation of markup on the retail price is consistent. In this edition, discussions and problems on markup will use the retail basis for all further calculations.

B. Markup Calculations Used In Buying Decisions

There are numerous purchase planning and merchandise pricing problems that face the retailer when buying goods for resale. To achieve maximum profits, the same estimated initial (original) markup percentage is not applied to all purchases. The astute merchandiser has set an ultimate goal, but realizes that situations arise that may cause deviations as individual purchases are made. Therefore, the manipulation of the markup and sales volume ultimately will provide the largest possible dollar profit. The calculations of the various buying and pricing situations that occur in merchandising are expressed in formulas and must be understood by all retail merchandisers.

1. Calculating Markup Percentage When Individual Cost and Individual Retail Are Known

CONCEPT:

$$\text{Markup \% (on retail)} = \frac{\$ \text{ Markup}}{\text{Retail}}$$

PROBLEM:

What is the markup percent on an item that costs $6.50 and is priced at $12.75?

SOLUTION:

Given:

Retail	= $12.75
Cost	= $ 6.50
Markup	= $12.75 (retail) – $6.50 (cost)
	= $ 6.25
Markup %	= $\dfrac{\$ \text{ 6.25 Markup}}{\$12.75 \text{ Retail}}$
Markup %	= 49%

When the individual cost and the individual retail do not change, but the number of purchased pieces varies, the markup percent is the same whether it is calculated for one piece or for the entire quantity purchased. The following calculations illustrate this principle:

PROBLEM:

What is the markup percentage on a purchase of 12 pieces that cost $6.50 each and are retailed at $12.75 each?

SOLUTION:

Total retail	= $153 (12 pieces × $12.75)
– Total cost	= – 78 (12 pieces × $6.50)
Dollar markup on entire purchase	= $ 75
Markup %	= $\dfrac{\$ \text{ 75 Markup}}{\$153 \text{ Retail}}$
MU% on Entire Purchase	= 49%

As a buyer evaluates a particular item, generally the markup percent on the individual piece will be used in making a determination of its salability. Once an order is placed, however, the markup percent will be calculated by using the entire quantity to be purchased.

2. Calculating Markup Percentage on a Group of Items with Varying Costs or Retail Prices

In the final analysis, purchases are evaluated on an overall basis. The following variations on markup formulas illustrate this same concept in that the markup percentage is calculated on the total amounts ordered — rather than on an individual basis — so that the total purchase can be evaluated.

CONCEPT:

$$\text{Markup \% on entire purchase} = \frac{\text{Total \$ markup}}{\text{Total retail}}$$

a. Calculating the Markup Percent When Writing Orders Placed for a Variety of Items and Prices. Currently, some stores are using computer generated purchase order systems and automated replenishment systems. Though the mechanics of order writing can vary, it is basic that the markup percent on the entire purchase requires a calculation of the total cost and total retail to determine the overall percentage on the purchase.

PROBLEM:

A buyer ordered 10 coats at a cost of $59.75 each to retail for $100 each, and 6 coats costing $79.75 each to retail for $150 each. What is the markup percentage on this entire purchase?

SOLUTION:

Total retail		= $1,900.00
↓		
10 pieces × $100	= $1,000.00	
6 pieces × 150	= + 900.00	
	$1,900.00	
− Total cost		= −1,076.00
↓		
10 pieces × $59.75	= $ 597.50	
6 pieces × 79.75	= + 478.50	
	$1,076.00	

Dollar markup on entire purchase = $ 824.00

Markup % = $ 824.00 Total dollar markup
$1,900.00 Total retail

Markup % on Entire Purchase = 43.4%

b. **Varying Retail Prices of Either a Classification or a Group that have the Same Cost.**

PROBLEM:

A buyer bought 150 handbags that cost $22.50 each. The buyer then retailed 50 pieces for $40 each, 75 pieces for $48 each, and the balance for $55 each. What is the markup percent on this purchase?

SOLUTION:

Total retail	= $6,975

$\downarrow$

50 pieces × $40	=	$2,000
75 pieces × $48	=	$3,600
25 pieces × $55	=	$1,375
$\downarrow$		$6,975

− Total cost

150 pieces × $22.50	= −3,375
Dollar markup on entire purchase	= $3,600

Markup % = $\dfrac{\$3,600}{\$6,975}$ $\dfrac{\text{Total \$ markup}}{\text{Total retail}}$

Markup % on Entire Purchase = 51.6%

c. **Offering Merchandise With Varying Costs at the Same Retail Price.**

PROBLEM:

A jewelry buyer has an unadvertised promotion on rings at a special price of $25 each. The group consists of 75 pieces that cost:

▶ 15 pieces @ $10.00 each.

▶ 40 pieces @ $12.50 each.

▶ 20 pieces @ $16.00 each.

What is the markup percent on this group?

SOLUTION:

Total retail	= $1,875

$\downarrow$

75 pieces × $25

− Total cost

$\downarrow$

15 pieces × $10.00	=	$ 150	
40 pieces × $12.50	=	500	
20 pieces × $16.00	= +	320	
		970	− 970
Dollar markup on entire group		= $	905

Markup % = $\dfrac{\$\ 905}{\$1,875}$ $\dfrac{\text{Total dollar markup}}{\text{Total retail}}$

Markup % on Entire Purchase = 48.3%

3. Calculating Retail When Cost and Desired Markup Percentage Are Known

Although cost prices can be quoted by individual or by-the-dozen prices, in pricing an item for retail the merchandiser "thinks" in terms of unit retail and so uses the cost per piece as a basis for calculating the retail per piece.

CONCEPT:

$$\text{Retail} = \frac{\text{Cost}}{100\% - \text{Markup \%}}$$

PROBLEM:

A manufacturer quotes a cost of $42 per dozen for a top; the markup that the buyer wants is 51.7%. At what retail price should the item be marked to obtain the desired markup?

SOLUTION:

Given: Cost   = $42 per doz.

         = $42 ÷ 12

         = $3.50 per piece

MU%   = 51.7%

Retail = $\dfrac{\$3.50 \ (\text{Cost})}{100\% - 51.7\%}$

         = $\dfrac{\$3.50}{48.3\%}$

         = $\dfrac{\$3.50}{.483}$

Retail = $7.25

4. Calculating Cost When Retail and Markup Percentage Are Known

In maintaining established price lines, a retailer who knows the required markup must be able to determine the maximum affordable price to pay for an item so that it can be sold profitably.

CONCEPT:

Cost = Retail × (100% - MU%)

PROBLEM:

A children's underwear buyer plans to retail petticoats for $10.95 with a 48% markup. What is the maximum price the buyer can pay for the petticoats to be sold at this price line?

SOLUTION:

Given: Retail = $10.95

MU %   = 48%

Cost   = $10.95 (Retail) × (100% − 48%)

         = $10.95 × 52% (Cost complement %)

         = $10.95 × .52 (Cost complement)

Cost   = $ 5.69

FIGURE 5 Purchase Order Form

DATE SHIPPED	TO BE SHIPPED FROM	DELIVERY DUE	F.O.B.	FREIGHT ALLOWED	TERMS
			Charges on shipments bought FOB New York or our premises must be prepaid		Dating Is From Date of Receipt of Goods

CLASS.	HOUSE # STYLE # LOT #	DESCRIPTION	1	2	3	4	5	6	7	8	9	QUAN.	UNIT COST	TOTAL COST	UNIT RETAIL	TOTAL RETAIL

TOTAL

M.U.%

THIS ORDER IS PLACED SUBJECT TO CONDITIONS ON BOTH SIDES

Signed _____   Countersigned _____
DEPT. MANAGER   MDSE. MANAGER

Date _____ 19 ___

Note to students: Photocopy and enlarge this form to use with Purchase Order Problems 1, 2, and 3.

▶ On Figure 5 (purchase order), p. 58, write and complete the orders for problems 1-3.

Purchase Order Problem #1

    (1.) Style #300 — 25 leather attachécases, costing $79.50 each, retailing for $169.99.

    (2.) Style #401 — 10 leather folders, costing $55 each, retailing for $125 each.

    (3.) Style #411 — 15 leather briefcases, costing $69.50 each, retailing for $150 each.

Write the order.

Purchase Order Problem #2

As the men's raincoat buyer, a manufacturer offers you a group of 225 pieces at one low price of $50 each. You decide to price this group as follows:

    (1.) 25 pieces, retailing for $75 each.

    (2.) 50 pieces, retailing for $90 each.

    (3.) 50 pieces, retailing for $125 each.

    (4.) 100 pieces, retailing for $138 each.

Write the order.

Purchase Order Problem #3

You purchase the following from a glove resource:

    (1.) Style #321 — 20 dozen children's wool gloves costing $66 per dozen.

    (2.) Style #492 — 30 dozen men's wool gloves costing $72 per dozen.

    (3.) Style #563 — 10 dozen women's wool novelty gloves costing $78 per dozen.

You decide to price them all at the same retail of $12 per pair.

Write the order.

7. A buyer purchases dresses at $42.50 each.

    (a.) At what minimum price must these dresses be marked to achieve a 52% markup?

    (b.) At what actual customary price line are the dresses most likely to be marked?

    (c.) What would be the percentage of markup if the dresses were priced at $79.50?

8. A buyer purchased:

    ▶ 500 nylon windbreakers costing $12 each, to sell at $22 each.

    ▶ 700 sweatpants costing $9 each, to sell at $17.50 each.

    ▶ 300 acrylic sweaters costing $16 each, to sell at $30 each.

    What is the markup percentage for this order?

9. Apply the following "typical" departmental markups to find the cost of an item that retails for $49.95 in each department:

    (a.) Millinery department — 59%.

    (b.) Electrical appliances — 27.8%.

    (c.) Junior sportswear — 55.8%.

    (d.) Gifts & clocks — 56.2%.

    What would cause the wide range of markup percentages among the various products?

10. A buyer of men's furnishings paid $96 per dozen for wool blend knit neckties. If the desired markup was 58%, what was the exact retail price per tie? What price line would the buyer probably use in ticketing the ties for the selling floor?

11. A buyer is interested in a group of closeout jackets costing $35 each. If the required markup is 53.6%, what would be the minimum retail price for each jacket? If the buyer had previously bought these jackets at $39.75 and retailed them for $79.50, what was this previous markup percentage?

12. For a holiday catalog, a buyer makes a special purchase of infant knit shirts that cost:

   ▶ 40 dozen at $60/dozen.
   ▶ 20 dozen a $72/dozen.
   ▶ 18 dozen at $66/dozen.

   If the shirts all retail at the same unit price of $11, what markup percent will be yielded?

13. A millinery buyer buys straw hats at $45 per dozen and sells them at $7.50 each. What markup percent has been achieved?

14. A toy buyer planned a special sale of dolls to retail at $25 each. If the overall markup on the purchase was 46%, what was the cost per doll?

15. Men's walking shorts that cost $132/dozen require a markup of 50%. What should be the retail price of each pair?

16. A suit that costs $67.50 has a markup of 62.5%. What should be the retail price of each suit?

17. A buyer purchased 50 assorted leather attaché cases that cost $79.50 each, and subsequently merchandised them as follows:

   ► 25 pieces to retail for $175 each.
   ► 10 pieces to retail for $150 each.
   ► 15 pieces to retail for $125 each.

Find the markup percentage on this purchase.

18. An outerwear buyer arranges a special purchase from a manufacturer who offers a group of 150 raincoats (with varying costs) at one low price of $25 each. The buyer decides to price this purchase as follows:

   ► 50 pieces to retail at $35.95 each.
   ► 50 pieces to retail at $45.00 each.
   ► 50 pieces to retail at $55.00 each.

What markup percentage is realized on this purchase?

19. A special promotion, consisting of 10 dozen novelty sweaters, is purchased at the cost of $240 per dozen. Determine the retail selling price of each sweater if the buyer wants to obtain a 52% markup.

20. After its arrival in the store, the buyer (in Problem 19) reviewed the shipment and decided to include this purchase in a store flyer featuring a group of sweaters retailing for $42 each. What markup percentage was achieved on this particular purchase?

21. If the shoe buyer wants to buy some leather sandals to retail for $49.50 each, and needs to obtain a markup of 53.4%, what is the most that these sandals can cost?

22. Calculate the markup percentage on the following order placed by the handbag department:

   ▶ 3½ dozen, costing $180 per dozen to retail for $32.50 each.

   ▶ 15 pieces, costing $37.50 each to retail for $79 each.

23. The jewelry buyer wants to spend $15,000 at retail. The department maintains a 56.7% markup. The buyer needs a minimum of 600 pairs of earrings to cover each branch with an adequate assortment.

   (a.) What should be the retail price per pair?

   (b.) What should be the cost per pair?

24. The sportswear buyer wants to spend $84,000 at retail. Orders for $18,500, at cost, have already been placed. If the planned markup percent to be achieved is 52.5%, how much, in dollars, does the buyer have left to spend at cost?

► III. REPRICING OF MERCHANDISE

The dynamic nature of merchandising makes the repricing of goods in retailing universal. Price adjustments are made to either increase or decrease the original retail price placed on merchandise. These changes in prices must be properly recorded to:

- ► Achieve an accurate book inventory figure used in the retail method of inventory.
- ► Plan initial (original) markup goals when pricing goods.
- ► Control and manage the amount taken in an attempt to merchandise at a profit.

The repricing of goods for sale is constant, the causes are numerous, and the skill required is considerable. It is rare that a retailer makes an upward adjustment to the retail price, and often a buyer is forced to make a downward retail price change on a significant portion of sales. In a profit and loss statement the downward differences frequently have a group heading of RETAIL REDUCTIONS that include:

- ► Markdowns.
- ► Employee discounts.
- ► Shortages[2].

Though all retail reductions have an impact on markup, they are, for the most part, considered a necessity. Their study is of major importance, for only with complete understanding of their use can the retailer turn them into a dynamic and advantageous merchandising tool.

A. Markdowns

The most common and most important type of price adjustment is technically called MARKDOWN (MD). It is the lowering or reducing of the original or previous retail price on one item or a group of items. For example, a sweater that was retailed for $25 when it was received in the store was reduced to $18 because it became soiled. This price adjustment is called a markdown because the retail value of the merchandise was lowered. The difference between the new selling price ($18) and the former price ($25), is the amount of $7. The amount by which the retail value has been lowered is called the markdown and is the meaningful figure to the merchandiser. The merchandiser expresses markdowns as a percentage of the net sales of all the goods during a period, month, or year. Frequently, the merchandiser will want to calculate the markdown percentage that is necessary to sell a group of items. When this occurs, the markdown is still expressed as a percentage of the net sales figure.

1. The Purpose of Markdowns

Markdowns are "a cure, not a curse". This merchandising tool can be used to good advantage if the retailer realizes the objectives of markdowns. The major aims of reductions are:

---

[2]Shortages will be discussed and calculated in Unit IV — Retail Method of Inventory.

- To stimulate the sale of merchandise to which customers are not responding satisfactorily.
- To attract customers to stores by offering "bargains."
- To meet competitive prices.
- To provide open-to-buy money to purchase new merchandise.

2. Causes of Markdowns

By analyzing all the possible causes of markdowns, a merchandiser can make an effort to minimize them. The most common causes (not in order of importance) are listed in the sections below:

- Buying errors, which include:
  Overbuying in quantities.
  Buying of wrong sizes.[3]
  Buying of poor styles, quality, materials, and colors.
  Poor timing in ordering goods.
  Receiving and accepting merchandise that has been shipped late.

- Pricing errors, which include:
  Poor timing of markdowns.
  Setting the initial price too high.
  Not being competitive in price for same goods.
  Deferring markdowns too long.
  Calculated risks of carrying "prestige" merchandise.

- Selling errors, which include:
  Poor stockkeeping.
  Careless handling that results in soiled and damaged goods.
  Failure to display merchandise properly or advantageously.
  Uninformed salespeople.

- Special sales from stock, which include:
  Off-price promotions.
  Multiple sales (e.g., 3 for $1.00).

- Broken assortments, remnants, etc.

- Necessary price adjustments.

- Remainders from special sales.

3. Timing of Markdowns

Accurate timing of markdowns can help reduce the amount of markdowns needed to sell the merchandise. It is suggested that merchandise be analyzed and reduced when:

- Merchandise becomes "slow selling".
- The customer demand is sufficient to sell the merchandise with a minimum price reduction.
- The consumer's interest in the merchandise in stock may diminish because of the appearance of a new fashion or product, or of a lower price.

---

[3]Due to unbalanced buying or accepting merchandise that is sized contrary to order.

4. The Amount of the Markdown

Judgment is required in determining the price at which items can be cleared quickly. The repricing of goods is a major factor in the control of markdowns. It is difficult to generalize on the amount of the markdown to be taken because the "right" price depends on:

- ▶ The reason for the reduction.
- ▶ The nature of the merchandise.
- ▶ The time of the selling season (the proper moment during the selling season).
- ▶ The quantity on hand.
- ▶ The original (initial) markup.

Because the purpose of a markdown is to sell the merchandise quickly, the size of the markdown must be large enough to produce the desired results. Some rules to be considered in repricing are:

- ▶ The first markdown should be sharp enough to move a considerable amount of the goods.
- ▶ Markdowns should be sufficiently large enough to be attractive to customers who rejected the merchandise at its original price.
- ▶ The price can be reduced sufficiently to appeal to the next price zone customer.
- ▶ Markdowns should not be so large as to invite customer suspicion.
- ▶ Small markdowns are ineffective.
- ▶ Small, successive markdowns may increase the total loss.

5. Markdown Calculations

a. Calculating the Dollar Markdown. To find the dollar amount of markdown taken when there is a group of items, it is customary to first find the difference per piece between the present and new retail prices, and then to determine the total cost of the markdown.

CONCEPT:
   Dollar markdown = Original retail price – New retail price

PROBLEM:
   A buyer reduces 93 scarves from $15 to $10. What is the total markdown in dollars?

SOLUTION:

Original or present retail	$15.00
– New retail	–10.00
Dollar markdown	$ 5.00 Per piece
Total Dollar Markdown =	93 pieces × $5.00 = $465

However, the retailer frequently advertises markdowns as a percentage off the current retail price; for example, "25% off the selling price".

CONCEPT:

Dollar markdown = Percent off × Present retail price

PROBLEM:

A store advertises 25% off on a group of 50 suits currently retailed at $100 each. What is the total dollar markdown?

SOLUTION:

Percentage off	=	25% × $100 Retail
Dollar Markdown	=	$25
	=	50 pieces × $25
Total Dollar Markdown	=	$1,250

b. Calculating the total dollar markdown when a second markdown is taken.
Because all merchandise doesn't sell despite the markdown taken, after a reasonable period of time, an additional reduction must be made to move the remaining goods of that style and/or group.

CONCEPT:

Total dollar markdown = First total dollar markdown + Second total dollar markdown

PROBLEM:

The men's clothing buyer had a group of 50 jackets, priced at $225 each that were selling very slowly. To stimulate sales of these jackets, the buyer reduced the group to $175. In a short time, at this price, 40 pieces sold. At a later date, the remaining pieces became shopworn and needed a further reduction to clear them from stock. The buyer reduced them to $100 each and at that price they all sold out. What was the total dollar markdown taken?

SOLUTION:

First Markdown on Group

Original retail price		$ 225	
– First markdown price		– 175	
Amt. of markdown per piece	= $	50	
× Number of jackets	×	50	
First dollar markdown	=	$2,500	$2,500

PLUS

Second Markdown on Group

First markdown price		$ 175	
– Second markdown price		– 100	
Amt. of markdown per piece	= $	75	
× Number of jackets	×	10	
Second dollar markdown	= $	750	+ 750
Total Dollar Markdown On This Group		=	$3,250

c. Calculating the planned dollar markdown when the markdown percent and net sales are known.
Once an estimated net sales figure for an accounting period has been projected, and the markdown percent allowed is known, the total amount of markdown in dollars can be determined. Because markdowns are taken in dollars off the retail price, a buyer's thinking and decisions in merchandising markdowns alternates, according to circumstances, from percentages to dollars and vice versa.

CONCEPT:

Planned dollar markdowns = Net sales × MD %

PROBLEM:

The sales in the men's footwear department were planned for $560,000 and the markdown percent was estimated at 5%. Find the dollar amount of markdowns that would be permitted.

SOLUTION:

Total planned dollar markdowns	= $560,000 Net sales
	× 5% Est. MD%
Total Planned Dollar Markdowns	= $ 28,000

▶ **For assignments for the previous section, see practice problems 25-29.**

d. Calculating the Markdown Percentage. Markdowns taken are expressed as a percent of the net sales for an accounting period. It is important to control and plan markdowns within a store. Markdown percentages can be calculated for an entire department, as shown in the following calculations:

CONCEPT:

$$MD \% = \frac{\$ MD}{\$ Net\ sales}$$

PROBLEM:

Department #33 had net sales for March of $5,000. The markdowns taken for March totaled $500. What was the markdown percent for March?

SOLUTION:

$$\textbf{MD \%} = \frac{\$500\ Markdown}{\$5,000\ Net\ sales} = 10\%$$

Also, markdown percentages can be used to evaluate a group of items and/or a vendor. The following shows this type of calculation:

CONCEPT:

$$MD \% = \frac{\$ MD}{Total\ dollar\ sales\ of\ group's\ final\ selling\ prices}$$

PROBLEM:

A buyer had a special selling price group of 100 faux snakeskin belts marked at $16 each. At the end of the season, the 15 pieces that were unsold were reduced to $10 each, and sold out immediately. What is the MD % on this purchase?

SOLUTION:

Step 1: Determine markdown per piece.
$16 to $10 = $6 each

Step 2: Determine markdown amount.
15 pieces reduced = 15 pieces × $6 = $90 Markdown

Step 3: Determine final selling price sales of group.

85 pcs × $16	=	$1,360
15 pcs × $10	=	+ 150
Sales of group	=	$1,510

$$MD \% = \frac{\$90\ Markdown}{\$1,510\ Sales\ of\ group}$$

MD % = 5.96% or 6%

FIGURE 6 POS Markdown Receipt

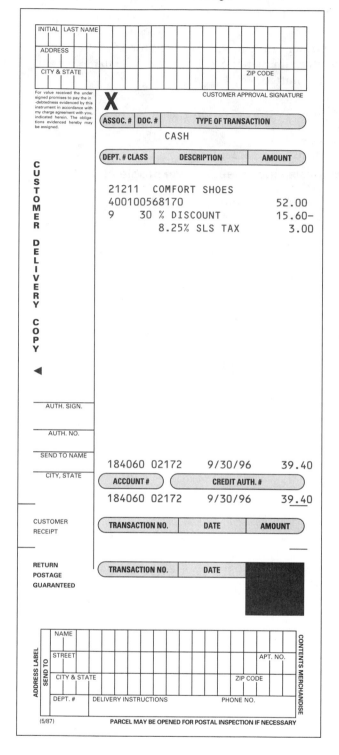

FIGURE 7 POS Credit Slip

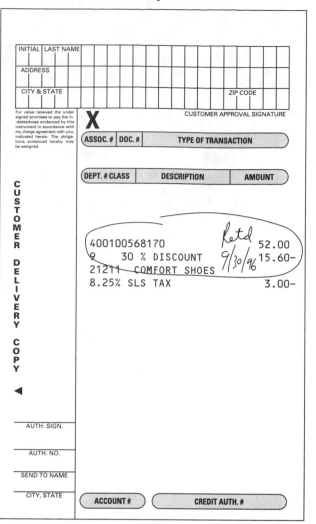

e. Calculating Markdown Cancellations. It is easier and more understandable to describe a MARKDOWN CANCELLATION than to define it. When the markdown price is raised to a higher retail price, it is considered a cancellation of a markdown. Cancellations may occur after special sales from stock, if the remaining merchandise is repriced upwards. The restoration of a markdown price to the former retail is a markdown cancellation.

Currently, markdown cancellations are much less common because most large stores electronically program temporary markdowns into the cash register. This means of recording the markdown as the reduced offering when an item is being sold is called a POINT-OF-SALE (POS) markdown. It eliminates the need to mark the goods with the reduced price, and, subsequently, to remark the goods to the original price. Figure 6 shows a sales slip recording the purchase of a pair of comfort shoes selling for $52 less 30%. The $15.60 is an example of a POS markdown.

Figure 7 is a credit slip indicating the return of "sale" merchandise and shows how the $15.60 markdown taken at the time of purchase was canceled. The amount of markdown taken on this simple transaction is zero because the $15.60 markdown recorded by the cash register on the purchase was neutralized by the return of the goods.

For those firms that are not yet using the POS system of recording and canceling markdowns, the following material, though not popular, is included to show the basic principles for the manual procedure. (Note: There are no review problems that illustrate this material.)

CONCEPT:

Markdown cancellation = Higher retail − Markdown price

PROBLEM:

After a one-day sale, a buyer marked up the remaining 12 pieces to the original price of $50, which had been reduced to $43 for the special event. What was the amount of the markdown cancellation?

SOLUTION:

Markdown cancellation = $50 Higher retail

−43 Markdown price

= $ 7 Difference per piece

= 12 Pieces × $7 Difference

Markdown Cancellation = $84

d. Calculating Net Markdown. When a reduction in price is made originally, it is the GROSS MARKDOWN. The difference between gross markdown and markdown cancellation is called the NET MARKDOWN figure or the amount of the permanent markdown.

CONCEPT:

Net markdown = Gross markdown – Markdown cancellation

PROBLEM:

A buyer reduces 75 pieces from $50 to $43 for a one-day sale. After the sale, the remaining 28 pieces were marked up to the original price and were sold out at that price.

(a.) What was the gross markdown in dollars?

(b.) What was the markdown cancellation?

(c.) Find the net markdown in dollars.

SOLUTION:

(a.): Determine gross markdown.

Gross markdown   = $50 – $43 = $7 per piece

               = 75 pcs. × $7 MD

Gross Markdown   = $525

(b.): Determine markdown cancellation.

Markdown cancellation   = $43 to $50 = $7 per piece

                    = 28 pcs. × $7 MD cancellation

Markdown Cancellation   = $196

(c.): Determine net markdown.

Gross markdown	= $525
– Markdown cancellation	= –196
Net Markdown	= $329

▶ **For assignments for the previous section, see practice problems 30-40.**

B. Employee Discounts

It is common practice for a retail store to give its employees a reduction off a retail price. This type of retail reduction is called an EMPLOYEE DISCOUNT. It was noted previously that this is another type of price adjustment that must be recorded because it lowers the value of merchandise. This reduction is generally stated as a percentage off the retail price (e.g., a 20% discount). It must be recorded by the statistical department of the accounting division for accuracy of the book inventory figure under the retail method of inventory. The cumulative amount of employees' discounts is usually shown on the profit and loss statement under the general heading of "retail reductions."

PROBLEM:

Store X grants its employees a 20% discount on all merchandise purchased. If a salesperson buys an item that retails for $16.95, what is the employee discount in dollars? What is the amount paid for the item?

SOLUTION:

Retail price of item	= $16.95
Discount allowed	= 20%
Employee discount	= $16.95 × 20%
	= $ 3.39
Employee's Price	= $16.95 – $3.39 = $13.56

C. Additional Markup

An ADDITIONAL MARKUP is a type of price adjustment that raises the price of merchandise already in stock. It increases the original retail price placed on the merchandise. It is taken after and in addition to the original (initial) markup. Although it is not a common type of adjustment, a provision must be made when it is necessary for an upward change of the retail prices of existing inventories. For example, an upward change of wholesale costs might necessitate the immediate upward revision of retail prices on existing inventory.

D. Markup Cancellation

MARKUP CANCELLATION is a downward price adjustment that offsets an original inflated markup to a markup that is more normal. It is used to adjust the markup on a purchase in accordance with the original intent, and is not to be used to manipulate stock values. Currently, it is controllable by law, which dictates the time period allowed from receipt to reduction of the purchase as well as the size of the markup resulting from this practice. It is justified when quoting "comparable value" prices used in promoting goods. For example, a resource offers a retailer gloves at lower than normal cost of $7 as opposed to the usual $10 cost. This item may be presently in stock and retailing for $22. During the permitted time, the gloves costing $7, which originally retailed at $22, are reduced to $14. This lowered price appears to the public as a markdown, yet results in a markup that is close to normal. The $8 difference between $22 and $14 is the markup cancellation.

FIGURE 8 Price Change

Run Date: 02/28/96 Run Time: 22:29:30			Price Change				Report No: Page: 12	
BUYER: 02			DEPT NO: 02					
P/C NO: 2331916 P/C TYPE: 01 MARKDOWN-PERMANENT, TOTAL STYLE					ORIGINATING LOCATION: 02 ENTRY DATE: 02-27-96			
EFFECTIVE DATE: 03/04/96				REASON: 02 REPRICING CLEARANCE				
DOCUMENT NO:								
APPLICABLE LOCATIONS 01 02 03 05 07 08 09 11 12 13 14 15 16 17 18 19 20 21 23 24 25 26 SPECIAL INSTRUCTIONS: REPRICING								
CLASS	VENDOR	STYLE	UPC	OLD RETAIL	NEW RETAIL	CURR TKT RETAIL	NEW TKT RETAIL	DESCRIPTION
41	1	6028	400100964255	11.00	7.00	11.00	7.00 ( )	LG CIRCULAR STRETC
41	1	6140	400101350996	6.00	4.00	6.00	4.00 ( )	CHANNEL CIRCLE STR
41	1	6143	400101365136	6.00	4.00	6.00	4.00 ( )	STRETCH MATTE META
41	1	9106	400101338239	11.00	7.00	11.00	7.00 ( )	1/2" MESH
41	1	9107	400101109839	14.00	9.00	14.00	9.00 ( )	MESH BELTS
41	1	9115	400101338246	13.00	9.00	13.00	9.00 ( )	ROLLED MESH
41	28	1602	400101479789	10.00	7.00	10.00	7.00 ( )	STUD BELT
41	28	1730	400101479796	10.00	7.00	10.00	7.00 ( )	STUD BELT
41	117	1305	400101271024	9.00	6.00	9.00	6.00 ( )	1 1/4" W/SILVER 3
41	117	1359	400101207726	5.00	3.00	5.00	3.00 ( )	3/4" CROCO SILVER

E. Price Change Procedures

An efficient system of reporting and recording price changes is important for three reasons, which are:

▶ Facilitation of the review of price adjustments.

▶ Accurate calculation of inventory records.

▶ Identification of shortages.

Figure 8 illustrates a computer printout of a buyer authorized price change.[4] It instructs the individual stores to change the retails on certain merchandise, where to change them (i.e., store location), if the change is permanent or temporary, what the new retails should be, and the reason for repricing. Presently, most systems do not require that the store count the merchandise, only remark it, because, financially the perpetual inventory system[5] has made the adjustment.

F. The Relationship of Repricing to Profit

Every price change has an impact on gross margin and net profits. Because the elimination of price adjustments in retailing is impossible, it is necessary that stores classify each type of adjustment separately so that they can be analyzed, planned, and controlled. Markdowns, the major type of price changes, reduce the retail price, causing a decrease in gross margin that is further reflected in a decrease and/or elimination of profit. In certain merchandise or classifications, the markdown risk is frequently anticipated. It is offset by planning a higher initial (original) markup. Unfortunately, profit cannot always be preplanned because there are forces in doing business that do not permit easy solutions to problems (e.g., non-acceptance of a newly introduced fashion or style, change in competition, etc.). However, it is important for the retailer to realize the effect of price adjustments on gross margin and profit as illustrated by the following example:

Initial retail value of new purchase	$105,000 =	100.0%
– Cost of goods sold	– 49,980 =	47.6%
Gross margin	$ 55,020 =	52.4%
– Operating expenses	– 45,990 =	43.8%
Net profit	$ 9,030 =	8.6%

The total retail reductions for the period are $5,000:

Net sales	$100,000 =	100.0%
– Cost of goods sold	– 49,980 =	50.0%
Gross margin	$ 50,020 =	50.0%
– Operating expenses	– 45,990 =	46.0%
Net Profit	$ 4,030 =	4.0%

---

[4]Price change forms are also shown in Unit IV.

[5]Perpetual inventory is discussed further in Unit IV.

25. A group of 75 wool sweaters priced at $55 each were selling slowly. Because it was late in the season, the buyer felt a markdown was needed to stimulate sales of these sweaters and reduced them to $42 each. What was the amount of markdown taken?

26. On Columbus Day, the suit buyer offered a "30% off" for an unadvertised special. 50 suits priced at $100 each were sold that day. How much in dollar markdowns were taken?

27. The children's buyer had a group of 150 ski jackets priced at $75 each. The weather was warm and sales were poor on this particular style. After 25 pieces sold, the buyer decided to reduce them to $68. At this price, 90 pieces sold. The remaining pieces were broken in size and color, so the buyer decided to clear them from stock and further reduced them to $58. What was the total dollar markdown taken?

28. The junior sportswear buyer planned sales for the month amounting to $700,000. The markdown percent was estimated at 12%. How much dollar markdown could be taken?

29. A dress buyer took markdowns in the month of July totaling $2,000. During this month, the net sales for the department were $18,000. If the original markdown plan for this month called for 14% markdowns, was the actual markdown over or under the original plan, and by how much in dollars?

30. For the month of November, a suit buyer took markdowns totaling $1,600. During this month, the net sales for the department amounted to $50,000. What was the markdown percent for this month?

31. In the month of July, a domestics buyer took markdowns totaling $1,800. During this month, the net sales for the department were $25,000. If the original markdown plan for this month called for 8% markdowns, was the actual markdown over or under the original plan, and by how much in dollars? What was the actual markdown percent?

32. The seasonal plan for a shoe department shows sales of $153,000, with planned markdowns of 6.4%.

   (a.)   What is the planned dollar amount of markdowns?

   (b.)   If the actual markdowns are $7,200, how does the real percentage compare with the plan? Is the actual percentage more or less than planned? By how much?

33. A buyer reduces 75 jackets from $39.50 to $32.95 for a sale. A short while after the sale, the remaining 30 jackets are reduced further to $27 and at that price they all sold out. Find the total markdown amount taken.

34. A coat department had sales of $60,000 for the month of November. During the month, the buyer reduced 50 coats from $60 to $40 and another 100 coats from $80 to $60.

   (a.) What was the total dollar markdown for the month?

   (b.) Find the markdown percentage for the month.

35. A buyer had 400 pocket calculators in stock at $25 each. They were on sale for $18 each for a one day special event. At the end of the day, 210 remained unsold and were sold eventually at their original price.

   (a.) What was the total markdown in dollars?

   (b.) What was the markdown percentage on this purchase?

36. A retailer purchased some terry slippers from a new resource. In evaluating the purchase, the records indicated that the total purchase of 360 pairs of slippers were originally retailed for $9.50 per pair. 290 pairs were sold at this price. The first markdown to $7 resulted in the sale of another 50 pairs. A second markdown to $5 was required on the balance of the merchandise to sell the remaining broken sizes. What were the retailer's total reductions in dollars?

37. For Columbus Day, the coat buyer advertised all suede coats at 35% off. The records indicated sales of:

   ▶ 50 pieces at original retail of $275 each.
   ▶ 30 pieces at original retail of $195 each.
   ▶ 10 pieces at original retail of $150 each.

   (a.) What was the total markdown in dollars?

   (b.) If the sales for October were planned at $80,270, with estimated markdowns of 9%, did the buyer have more markdown money to spend for the month? If so, how much? If not, why?

38. If the markdowns in problem 37 were the only ones taken for the month, what was the markdown percentage for the month of October?

39. A buyer bought 300 leather belts that were retailed at $18 each. At the end of the season, the 25 unsold belts were cleared at $10 each and sold immediately. What was the markdown percentage on this purchase?

40. The lingerie department's estimated sales for the entire Spring Season were $750,000, with anticipated markdowns of 11.5%. At the end of the season, the buyer had taken $67,500 in markdowns. Were the actual markdowns taken more or less than planned? By how much in dollars and percent?

41. A buyer wants to sell snowsuits at $39.95 each. The markup is 48.5%. How much can the buyer afford to pay for each snowsuit?

42. If a manufacturer suggests a retail price of $12.50 for a line of tote bags costing $84.00 per dozen, what is the suggested markup percentage?

43. A manufacturer offered woven belts at $54 per dozen. If a buyer were to take a 47.5% markup, at what retail price would each belt sell?

44. A buyer purchased 1,500 pairs of men's lounging pajamas at $10 each for a Father's Day sale. The advertised price was $20 retail and 1,000 pairs were sold at this price. After Father's Day, the remaining pieces of this special purchase were integrated into stock at $25 each. Between July 1 and September 1, all but 50 pairs were sold at this price. The remainder was then cut to $12 each, and all sold at that price except 10 pairs that were too damaged and soiled to sell at any price, and therefore were salvaged.[6] What is the markup percentage on this purchase?

45. Find the markup percentage on the following handbag purchases:

   ▶ 36 bags costing $22, with a retail of $45.
   ▶ 24 bags costing $28, with a retail of $55.
   ▶ 12 bags costing $41, with a retail of $75.
   ▶ 12 bags costing $53, with a retail of $95.

46. Down quilts costing $82 per unit were promoted at an advertised retail price of $160 each. Of the 800 pieces purchased, 600 sold at full price, and another 140 at a one-day special mid-season sale of $125 each. 35 more were cleared at $90 in the markdown corner. The 25 soiled counter samples were salvaged. What markup percentage was yielded on this purchase?

---

[6]Goods unfit for resale and, therefore, were reduced to zero value.

47. A purchase of 350 exercise bikes was made at a cost of $106 each and was unit priced at $200 each. They sold well at the regular price for a time, but ultimately, the last 45 pieces were reduced and sold at $155 each.

    (a.)   What was the markup percentage on this purchase when the order was written?

    (b.)   What was the markdown percentage on the sale of the 350 machines?

48. A job lot of 650 bow ties costs $1,950. If 400 are priced at $6 and the remainder at $7.50, what markup percentage is achieved?

49. A junior sportswear department took the following markdowns for August:

    ▶ 160 tops from $28 to $21.

    ▶ 230 tops from $25 to $17.

    ▶ 170 tops from $20 to $14.

Gross sales for the month in this department were $76,500 and customer returns totaled 6%. What were the departmental markdown percentages for the month?

50. For a Veteran's Day sale, a buyer reduced 95 coats from $95 each to $69 each. The following day the buyer marked 59 additional coats from $125 to $95. What was the total dollar markdowns taken?

51. A buyer had 96 coats in stock at a retail price of $85 each. For a special sale, they were marked down to $55. The markdowns for this month were planned at 6%, with sales for the month estimated at $90,000. After this event, did the buyer have more markdown money to spend? If so, how much more? If not, why?

52. A lamp retailing for $97 was advertised at 30% off. What was the new selling price of the lamp after the markdown was taken?

53. In May, markdowns taken in a children's department were:

▶ 60 jackets from $18 to $15.
▶ 30 sweatshirts from $15 to $12.
▶ 70 T-shirts from $10 to $7.

Gross sales for the month in this department were $85,000 and customer returns totaled 6%. What was the departmental markdown percentage for the month?

54. The seasonal plan for a swimwear department showed planned sales of $650,000 and planned markdowns of 18%.

(a.) What was the amount of planned markdown dollars?

(b.) If the markdowns amounted to $120,000, was the actual percentage more or less than the planned percent? By how much in percent?

55. For Valentine's Day, the candy department buyer purchased a special offering of 150 heart-shaped boxes of candy that cost $6 each. There were two assortments, and the buyer decided to retail 75 boxes at $15 each and 75 boxes at $10 each. On February 15, the remaining $15 assortments were reduced to $9, taking total markdowns of $144, and the balance of the $10 assortment were reduced to $6 each, with these markdowns amounting to $72. Six of the $15 hearts that didn't sell at $9 were reduced to $5 and were cleared out. Four of the $10 hearts were reduced to $3 and sold out at that price. The sales of the department for February were $9,750.

    (a.) Determine the markup percentage on the original order.

    (b.) What was the markdown percentage on this purchase?

    (c.) What was the markdown percentage for the month if these were the only markdowns taken?

    (d.) What was the markup percentage achieved on this purchase?

56. "Markdowns are a normal and positive element of merchandise planning. The markdown factor may be used advantageously by a buyer in an effective departmental operation." Explain in detail.

Pricing in Merchandising

At the end of October, a key resource offered a department store buyer a group of acrylic knit tops that are promotionally priced at $96 per dozen. The resource, who owns 60 dozen pieces, quoted this low price providing the purchaser bought the entire group. These tops are comparable in quality and styling to similar merchandise purchased from this vendor at $144 per dozen that sold well at $25 each earlier in the season with the normal department markup. The group now being offered by the resource consists of three different styles currently in demand. Group A has 30 dozen pieces, Group B consists of 20 dozen pieces, and Group C is 10 dozen pieces. As a combined offering, the sizes are well balanced and the colors are "fashion" for this season. Group A is a "staple" in styling, Group B reflects this season's "hot seller", and Group C is considered "avant garde" in both color and silhouette. Because sales are sluggish, the buyer wants to stimulate selling for Election Day by retailing these tops with only a 50% markup. By taking a lower than normal markup, the buyer hopes this purchase will result in additional sales and be a total "sellout".

As the buyer reviews the stock condition of the department, it seems the inventory is higher than planned and the report shows that:

(1.) 55% of the inventory is current and salable.

(2.) 150 pieces remain from a recent promotion, currently priced at $19 each, and are broken in size and color.

(3.) A group of 75 pieces, all one style, priced at $22.50 each, are selling very slowly.

(4.) 15 pieces are badly soiled and are marked $25 each.

The buyer also noted that there is $1,000 of markdown money available in November.

The buyer feels the purchase would be ultimately advantageous because increased departmental activity will lead to improved sales. To assure this high volume day, the proper pricing of this purchase, plus an accurate count of the current inventory is essential.

To make the correct decisions, the following questions must be answered:

(a.) If this group purchase were to be retailed at the normal markup, what would be the retail per piece? What degree of excitement would this produce by quoting a higher "comparable value"?

(b.) Determine the markup percentage achieved if the three groups are retailed as follows:

▶ Group A at $16 each.
▶ Group B at $18 each.
▶ Group C at $20 each.

The last group seems to represent particularly good value.

(c.) Should the available November markdown money ($1,000) be used now to clear the present inventory, by reducing the three groups all to $16 and using their comparable value prices? Or should the buyer defer these reductions (which seem inevitable), until the store-wide clearance in December?

Decide on a course of action the buyer should pursue and mathematically justify your suggestions as you apply the principles in pricing and repricing.

UNIT III
The Relationship of Markup to Profit

OBJECTIVES

▶ Identification of the types of markups and the description of how each is used in making merchandise decisions.

▶ Calculation of different types of markups, including:
Initial markups
Cumulative markups
Maintained markups

▶ Recognition of the need for balancing markups among different items of merchandise to achieve merchandising goals.

▶ Balance of markups for diverse given situations.

▶ Recognition and understanding of both the dollar and percentage amounts of markup that are needed to evaluate the impact of merchandising decisions.

KEY TERMS

aggregate markup
alteration costs
cumulative markup
final selling price
gross margin

initial or original markup
initial or original retail
maintained markup
reductions
vendor analysis

he significance of pricing and repricing as well as the considerations that influence the retailer when setting retail prices were examined in Unit II. From that discussion, it is understood that the retail price-placed on merchandise should be based on the markup deemed appropriate by the merchandiser. Additionally, it should be recognized that markup has a relationship to and an affect on the amount of:

- ▶ Sales volume it can generate.
- ▶ Markdowns it can cause.
- ▶ Gross margin it can achieve.
- ▶ Profits it can produce.

For these reasons, the markup in a store or a department must be estimated, always controlled, and if necessary altered when conditions require it. Markup, as defined in Unit II, is simply the difference between retail prices placed on merchandise, and the cost of this merchandise. As already discussed, the basic markup calculations in both dollars and percentages are used in buying decisions as they relate to a single item or a group of items. The buyer is responsible for an entire department, and is involved with the markups planned and obtained for the whole department. These markups are constantly monitored because the desired gross margin depends on how successfully they are achieved.

To better understand the various markups that are indigenous to the retailing business, the term "retail price" must be clarified. The first price placed on merchandise for resale is the ORIGINAL or INITIAL RETAIL. The price received when an item sells, (which may be different), is the FINAL SELLING PRICE. The original, initial, or first retail placed on merchandise is the retail price the retailer hopes to realize, and the final selling price is the price received when it eventually sells, (e.g., a raincoat costing $55 is retailed at $120 when received in the store and eventually sells for $100.) The initial retail price is often higher than the price at which the merchandise ultimately sells, because price reductions may be necessary before the goods are sold.

▶ I. TYPES OF MARKUP

A. Initial Markup

INITIAL MARKUP[1] is the difference between the billed cost of merchandise and the original or first retail price placed on a given item or group of items. When freight charges are known, they are added to the billed cost in calculating initial markup. In the determination of an initial markup there are vital factors and common practices that must be considered and cannot be ignored. It is essential for a retailer to:

- ▶ Plan, in advance, an initial markup on merchandise when received in the store to assure a favorable gross margin figure that is large enough to satisfy expenses and provide a reasonable profit.
- ▶ Forecast an initial markup percentage that will be required on all the merchandise handled during a particular period (i.e., a six-month season), because it is virtually impossible to have available all the data necessary for the calculation of each purchase.

---

[1]Initial markup is also known as "markon" and the "original markup."

- Recognize that the gross margin figure can fluctuate because it is the result of many merchandising decisions.

- Understand that the initial markup placed on new purchases will be reduced by markdowns, shortages, and employee discounts.

- Be aware of and act upon cash discounts offered retailers in the purchase of goods, as well as estimated alteration costs.

For a department, an initial markup that is planned on a seasonal and/or annual basis can be expressed in either dollar or percentage figures and usually is calculated for both. In most large retail stores, the buyer and/or divisional merchandise manager (under the guidance and supervision of the control division), establishes this initial markup as a guide to check against aggregate markups obtained in pricing new merchandise purchases. Additionally, it is essential to know that the initial markup planned for a season must take into consideration the markups on the new purchases and the effect they have on the markup of the merchandise inherited from the last season, because the gross margin figure for the period depends on the markups achieved on all the inventory available for sale. Consequently, in planning an initial markup, probable expenses, the profit goal, probable reductions, estimated alteration costs, and anticipated sales must be projected first.

An initial markup is expressed as a percentage of the aggregate original retail price placed on the merchandise, not on the price at which the merchandise sold. The price at which the merchandise sold is expressed as sales plus reductions. For example, if sales are planned at $1,000,000 and all retail reductions at $150,000, the goods that eventually sell at $1,000,000 must be introduced into stock at $1,150,000 with the desired markup expressed as a percentage of this total. This seasonal planned markup percentage can be calculated and the concept can be expressed by a formula.

1. Establishing an Initial Markup

To better appreciate the establishment of an initial markup, examine the methods commonly used to institute a seasonal markup and the calculations (in both dollars and percentages) as they appear in the chart and steps listed below. (The figures listed in the chart are for illustration only, they do not represent actual retail figures.)

GIVEN:

	Dollars	Percentages
Planned sales	$1,000,000	100.0%
Estimated expenses	393,000	39.3%
Price reductions	150,000	15.0%
Profit	50,000	5.0%

METHOD:

(1.) Forecast the total sales for the season or the year ($1,000,000).

(2.) Estimate the required expenses ($393,000) and price reductions ($150,000) needed to reach the sales plan.

(3.) Set a profit goal — operating profit or department contribution ($50,000).

(4.) Add the estimated expenses ($393,000) to the price reductions ($150,000) and profit ($50,000) to determine the dollar markup ($593,000).

	Dollars	Percentages
Estimated expenses	$393,000	39.3%
+ Price reductions	+150,000	+15.0%
+ Profit	+ 50,000	+ 5.0%
Dollar Markup	$593,000	59.3%

(5.) Determine the original retail price ($1,150,000) by adding the planned reductions ($150,000) to the planned sales ($1,000,000).

	Dollars	Percentages
Planned sales	$1,000,000	100.0%
+ Price reductions	+ 150,000	+ 15.0%
Original Retail Price	$1,150,000	115.0%

(6.) Compute the markup percent (51.6%) by dividing the dollar markup ($593,000) found in step 4 by the original price ($1,150,000).

	Dollars	Percentages
Markup	$ 593,000	59.3%
÷ Orig. retail price	÷1,150,000	÷ 115.0%
Initial Markup	51.6%	51.6%

2. Calculating the Initial Markup

The formula for calculating an initial markup percent on goods purchased is a useful guide that shows the amount of markup necessary to achieve a desired profit. As discussed in Units I and II, this markup should be adequate to cover the factors of expenses, reductions, and profits. These projections can be based on past history or calculated mathematically on the basis of anticipated sales, reductions, and expenses. The determination of an initial markup percent can involve spelling out each factor (i.e., expenses + profits + markdowns + stock shortages + employee discounts) OR the equation can be simplified (i.e., expenses + profits = gross margin and markdowns + stock shortages + employee discounts = reductions). In the planning process, the gross margin and reduction figures are more significant than the analysis of expenses, shortages, etc., and consequently, the basic equation more often used is this second, simplified version.

a. Finding Initial MU Percentage When Gross Margin Percentage and Reduction Percentage Are Known.

CONCEPT:

$$\text{Initial markup \%} = \frac{\text{Gross margin \% + Reductions \%}}{100\% \text{ (Sales) + Reductions \%}}$$

PROBLEM:

A store has a gross margin of 42.3% (39.3% expenses + 3% profit) and reductions (markdowns, shortages, and employee discounts) of 15%. What is the initial markup percentage?

SOLUTION:

Initial MU % = 42.3% G.M. (39.3% Exp. + 3.0% Profit)
+15.0% Red. (MD's, Sh., Empl.Disc.)

$$\frac{100\% \text{ Sales} + 15\% \text{ Red.}}{}$$

$$= \frac{57.3\%}{115.0\%}$$

Initial Markup % = 49.8%

b. Finding Initial MU Percentage When Gross Margin and Reductions in Dollars Are Known.

CONCEPT:

$$\text{Initial markup \%} = \frac{\$ \text{ Gross margin} + \$ \text{ Reductions}}{\$ \text{ Sales} + \$ \text{ Reductions}}$$

PROBLEM:

A store plans sales of $1,000,000, reductions of $150,000, and requires a gross margin of $423,000 (i.e., expenses $393,000, profit $30,000). What should be the initial markup percentage?

SOLUTION:

$$\text{Initial MU \%} = \frac{\$ \ 423,000 \ (\text{G.M.}) + \$150,000 \ (\text{Red.})}{\$1,000,000 \ (\text{Sales}) + \$150,000 \ (\text{Red.})}$$

$$= \frac{\$ \ 573,000}{\$1,150,000}$$

Initial MU % = 49.8%

c. Finding Initial MU Percentage When Cash Discounts and Alteration Costs Are Known.
As discussed in Unit I, the cost of goods sold is adjusted by alteration and workroom costs and cash discounts. Accordingly, these must be calculated in the markup equation if and when they exist. Because markup must cover alteration and workroom costs, they are added to the gross margin. By the same token, because cash discounts reduce the cost of goods sold, which affects the final markup obtained, the cash discounts are subtracted from the gross margin. These factors are essential in determining retailing markups.

CONCEPT:

$$\text{Initial markup \%} = \frac{\text{G.M. \%} + \text{Alt. cost \%} - \text{Cash disc. earned \%} + \text{Red. \%}}{\text{Sales} \ (100\%) + \text{Red. \%}}$$

PROBLEM:

The desired gross margin of a store is 42.3%, the reductions are 15%; the cash discounts earned are 3%, and the alteration costs are 2%.

SOLUTION:

Initial Markup % = 42.3% G.M. + 2% Alt. costs
−3.0% Cash disc. + 15% Red.

$$\frac{100\% \ (\text{Sales}) + 15\% \ \text{Red.}}{}$$

$$= \frac{56.3\%}{115.0\%}$$

$$= 48.96\%$$

Initial MU % = 49%

▶ **For assignments for the previous section, see practice problems 1-10.**

B. Cumulative Markup

Even though an initial markup may be calculated for individual items, the initial markup on merchandise received is more commonly reported for season-to-date, or for the year. Over any such extended period, an initial markup is called CUMULATIVE MARKUP, which is the markup percentage figure generally used by retailers to compare the merchandising performance and information with other stores. Cumulative markup is the markup percentage achieved on all goods available for sale from the beginning of a given period. It is an average markup because it is the markup percentage obtained on the accumulated inventory at the beginning of the given period, plus the markup of all the new purchases received season-to-date. The cumulative markon in dollars equals the difference between the invoiced cost of the merchandise (including transportation) before cash discounts have been adjusted and the cumulative original retail prices of all merchandise handled (opening inventory + net purchases) during a given period of time. Markdowns do not enter into the calculation of the cumulative markup percentage. The concept is simply stated by saying:

CONCEPT:

$$\text{Cumulative markup percent} = \frac{\text{Cumulative markup dollars}}{\text{Cumulative retail dollars}}$$

PROBLEM:

On February 1, a boys' clothing department has an opening inventory of $200,000 at retail with markup of 49.0%. On July 31, the new purchases season-to-date amounted to $1,350,000 at retail with a 49.9% markup. Find the cumulative markup percent achieved.

SOLUTION:

Given information is in **bold**, and the other items are calculated from the basic markup formulas; each calculation is identified with the steps of the procedure.

	Cost	Retail	MU%
Opening inventory	$102,000	**$ 200,000**	**49.0%**
+ Purchases STD	+676,350	**+1,350,000**	**49.9%**
Total mdse. handled	$778,350	$1,550,000	49.8%

Step 1: Find the cost value of the retail opening inventory.

C. = R. × (100% – MU%)

C. = **$200,000** (100% – **49.0%**)

C. = **$200,000 × 51%**

C. = $102,000

Step 2: Find the total cost value of all purchases season- to-date.

C. = R. × (100% – MU%)

C. = **$1,350,000** (100% – **49.9%**)

C. = **$1,350,000 × 50.1%**

C. = $676,350

Step 3: Find the total retail value of total merchandise handled.

$200,000 + $1,350,000 = $1,550,000

Step 4: Find the total cost value of total merchandise handled.

$102,000 + $676,350 = $778,350

Step 5: Find the cumulative markup on total merchandise handled.

$1,550,000 Cumulative retail

− 778,350 Cumulative cost

$ 771,650 Cumulative markup

$771,650 ÷ $1,550,000 = 49.8%

Cumulative Markup % = 49.8%

The new purchases required a 49.9% markup to achieve a 49.8% cumulative markup, because the merchandise of the opening inventory came into the period with only a 49% markup.

At this point of understanding the relationship between initial and cumulative markups, the realization must be made that both of these markups are on merchandise purchased. Initial markup percent refers to those markups obtained when pricing new merchandise purchases. Cumulative markup percentage is the amount of markup achieved on all the merchandise available for sale, whether it is new purchases or stock-on-hand at the beginning of the period. The cumulative markup percentage is the initial markup percentage calculated from the beginning of the season to any given later date (e.g., the end of the season or year).

▶ **For assignments for the previous section, see practice problems 11-15.**

C. Maintained Markup

Before maintained markup is explained, it is essential to know that while there is a relationship between gross margin and a maintained markup, they are not identical. Gross margin is the difference between net sales and the cost of merchandise sold, adjusted by subtracting cash discounts and adding alteration/workroom costs (i.e., total cost of merchandise sold). MAINTAINED MARKUP (MMU) is the difference between net sales and the cost of merchandise sold without the credits of cash discounts and without adding alteration/workroom costs (i.e., gross cost of merchandise sold). Consequently, if these differences are not considered in calculations, the gross margin and maintained markup figures would be the same. For both, the net sales figure reflects the final selling prices received for the goods sold, and the margin actually realized when the goods are sold. They are both markup on sales.

The relationship between gross margin and maintained markup is revealed clearly in the following example:

Net sales		$1,000,000
Cost of merchandise sold		
Opening inventory	$ 220,000	
+ New purchases	+ 645,000	
+ Inward freight	+ 14,000	
Total merchandise handled	$ 879,000	
− Closing inventory	− 280,000	
Gross Cost of Merchandise	$ 599,000	
− Cash discounts earned	− 32,000	
Net cost of merchandise sold	$ 567,000	
+ Alteration/workroom costs	+ 10,000	
Total Cost of Merchandise Sold		− 577,000
Gross Margin		$ 423,000

CONCEPT:

G.M. % = Net sales − Total cost of mdse. sold

= $1,000,000 (Sales) − $577,000 (Total cost of mdse. sold)

$$= \frac{\$ 423,000 \text{ G.M.}}{\$1,000,000 \text{ Net sales}}$$

G.M. % = 42.3%

While

CONCEPT:

MMU % = Net sales − Gross cost of mdse. sold

= $1,000,000 Net sales

− 599,000 (Gross cost of mdse. sold)

$$= \frac{\$ 401,000 \text{ Maintained margin}}{\$1,000,000 \text{ Net sales}}$$

MU% = 40.1%

Maintained markup is the markup actually achieved on the sale of the merchandise and generally is calculated for a department or, when desired, for a VENDOR ANALYSIS. A vendor analysis is an investigation of the profitability of each vendor's products that are sold by a retailer. It is not customary to plan a maintained markup because it is the result of merchandising activities. The initial markup must be planned in advance so that the desired final or maintained markup goal is achieved. Additionally, it should be noted that the maintained markup must be large enough to cover expenses and provide a profit while, as previously explained, the initial markup must be high enough to anticipate and also cover all possible retail reductions (i.e., markdowns, shortages, and employee discounts). However, once the initial markup and reductions are planned, the probable maintained markup can be projected.

1. Finding Maintained Markup When Initial Markup and Reductions Are Known

CONCEPT:

MMU % = Init. MU % – Red. % × (100% – Init. MU %)

PROBLEM:

A department planned an initial markup of 49.8% and reductions of 15%. What is the maintained markup?

SOLUTION:

Maintained MU %   = 49.8% – 15% × (100% – 49.8)

                = 49.8% – 15% × 50.2%

                = 49.8% – 7.53%

Maintained MU %   = 42.27 or 42.3%

2. Finding Reductions When Initial Markup and Maintained Markup Are Known

CONCEPT:

$$\text{Reduction \%} = \frac{\text{Initial markup \% – MMU \%}}{\text{Sales (100\%) – Initial markup \%}}$$

PROBLEM:

The department planned an initial markup of 49.8% and wanted to achieve a maintained markup of 42.3%. What are the reductions?

SOLUTION:

$$\text{Reduction \%} = \frac{49.8\% \text{ Init. MU} – 42.27 \text{ MMU \%}}{\text{Sales (100\%) – 49.8\% Init. MU}}$$

$$= \frac{7.53\%}{50.2\%}$$

Reduction %   = 15.0%

The same factors have been used deliberately to show the calculations of both initial and maintained markups, so that the relationship can be fully appreciated. The maintained markup concepts can also be expressed in dollars and the procedure for calculation is identical. A particular merchandising situation, however, may require the focus to be on dollars rather than percentages, (e.g., merchandising fast-turnover goods).

▶ **For assignments for the previous section, see practice problems 16-22.**

NOTES

1. A lingerie buyer determines that the department has net sales of $750,000, expenses of $315,000, and total reductions of $75,000. This buyer also wants to attain a net contribution of 4.5%. Find the initial markup percentage.

2. A department shows a gross margin of 41% (37.9% expenses + 3.1% profit) and lists reductions (i.e., markdowns, shortages, and employee discounts) of 13%. What is the initial markup percentage?

3. A chain of specialty shops plans sales of $1,500,000 and reductions of $260,000. It needs a gross margin of $605,000 (i.e., expenses $550,000, profit $55,000). What should be the initial markup percentage?

4. In a small leather goods department, the markdowns, including employee discounts, were 18.7%, stock shortage was 3.8%, and the gross margin was 46.6%. Determine the initial markup percentage.

5. A retailer in a boutique jewelry store has estimated expenses of 39%, markdowns at 15%, and stock shortage at 6.3%. A profit of 4% is desired. Calculate the initial markup percentage required.

6. The intimate apparel buyer wants to determine an initial markup percent for the robe classification, from the following data:

    ▶ Gross margin = 44.0%.

    ▶ Markdowns = 31.4%.

    ▶ Shortages = .6%.

Calculate the initial markup percentage.

7. A gross margin of 49.4% is targeted by a gift department. Reductions are 16% and cash discounts earned are 4%, with alteration costs of 1%. Find the initial markup percentage.

8. A buyer plans, for the period, net sales of $1,500,000 with a 49% gross margin. The acceptable markdowns are 12%, with employee discounts at 1%, and planned shortages of 2.5%. Cash discounts to be earned are estimated at 6%, and alteration costs are .5%. What initial markup percentage is needed to achieve the desired results?

9. Determine the initial markup percent from the following report:
   - ► Gross margin   = 46.8%.
   - ► Markdowns   = 10.0%.
   - ► Employee discounts   = 1.5%.
   - ► Cash discounts to be earned = 5.0%.
   - ► Alteration costs   = .5%.

10. The petite sports wear buyer plans seasonal net sales of $2,000,000 with a gross margin of 48%, markdowns (including employee discounts) estimated at 32.9%, planned shortages to be 2.1%, cash discounts to be earned planned at 6%, and alteration costs at 1%. Calculate the planned initial markup percentage.

11. A hosiery buyer has an opening stock figure of $180,000, at retail, which carries a 52% markup. On March 31, new purchases since the start of the period were $990,000 at retail, carrying a 54% markup. Find the cumulative markup percent on merchandise handled in this department to date.

12. The slipper department showed an opening inventory of $80,000 at retail with a markup of 48%. The purchases for that month amounted to $30,000 at cost, which were marked in at 52%. The initial markup planned for this department was 50.5%.

    (a.) Determine the season-to-date cumulative markup percent for the department.

    (b.) Was the department on target with its markup? If not, what factors caused a deviation?

13. A glove department had an opening inventory of $95,000 at retail, with a 55.8% markup. Purchases during November were $64,000 at cost, and $142,000 at retail. Determine:

   (a.) The cumulative markup percent.

   (b.) The markup percent on the new purchases.

14. During the fall season, a retailer determined that the total amount of merchandise required next season to meet the planned sales for that period was $360,000, at retail, with an initial markup goal of 52%. At the beginning of the next season, the merchandise on hand (opening inventory) came to $80,000 at retail, with a cumulative markup of 49% on these goods. For the coming season, what initial markup percent does the buyer need to achieve on any new purchases?

15. In preparation for a foreign buying trip, a buyer determines that a 55.4% markup is required on purchases that will amount to $560,000 at retail. While on this trip the following purchases are made:

	Cost	Retail
Resource A	$20,000	$ 45,000
Resource B	$50,000	$125,000
Resource C	$70,000	$170,000

What markup percent is needed on the balance of the purchases?

16. A men's shop with an initial markup of 53% had markdowns of 12%, employee discounts of 2.5%, and shortages of 1.5%. What was the maintained markup percent?

17. A sporting goods store has an initial markup of 44.5%. The expenses are 31%, markdowns are 12%, cost of assembling bicycles, etc., (i.e., workroom costs) are 6%, and shortages are 1%. What was the maintained markup percent?

18. The Closet Shop had the following operational results:

> ▶ Net sales = $125,000.
> ▶ Billed cost of new purchases = $64,000.
> ▶ Inward freight = $1,000.
> ▶ Alteration costs = $2,000.
> ▶ Cash discounts = $7,000.

Find:

(a.) The maintained markup in dollars and percentages.
(b.) The gross margin in dollars and percentages.

19. The jewelry department has an initial markup of 55.6%, with total retail reductions of 15%. There are no alteration costs or cash discounts. What is the maintained markup percentage and the gross margin percentage?

20. A children's store had sales of $500,000, with markdowns of 14.5% and shortages of 2%. The initial markup was 51.7%. What was the maintained markup percentage?

21. The T-Shirt department buyer determined that the department's initial markup should be 51.5%. The buyer also wanted to attain a maintained markup of 45%. Under this plan, what reductions (in percentages) would be allowed?

22. A boutique had planned a gross margin of 49%, with total retail reductions of 18%. At the end of the period, the maintained markup attained was actually 48.8%.

(a.) Find the initial markup percent needed to achieve the planned gross margin of 49%, with total retail reductions of 18%.

(b.) Find the actual amount of markdowns (in percentages) taken.

(c.) Prove your calculations and explain your findings. What may have caused the maintained markup percent to vary from the planned gross margin target?

► II. AVERAGING OR BALANCING MARKUP

A buyer's skill is truly tested by the ability not only to buy "the right goods, at the right time, in the right amounts", but, additionally, a buyer must also be able to achieve a predetermined markup percentage over a season or year. Failure to reach this goal can have an adverse effect on profits, and sometimes can determine whether or not a profit is generated. Ultimately, as short-term buying decisions are made, the merchandiser must know how any deviations from the markup objective will effect the long-term performance.

In the actual pricing of goods in retailing, it is seldom possible to obtain the same markup on all lines of merchandise, all classifications, all price lines, or all items carried within a particular department. In the real world of retailing, deviations from the planned seasonal markup will occur. These differences can be caused by competition, variations in special promotional merchandise offered by resources, special buying arrangements, private brands, imports, and other factors that rely on the buyer's judgment. Accordingly, to realize the planned markup necessary for a profitable operation, below-average markups should be balanced by above-average markups. Averaging markups means adjusting the proportions of goods purchased at different markups, to achieve the desired aggregate markup, either for an individual purchase or for a certain period. In merchandising the purchases, the buyer builds a "cushion" or lowers the markup depending on the situation. This practice is prevalent in the fashion and fashion-related industries. The appropriate cumulative markup originates from effective averaging of many purchases.

When solving problems that illustrate how markups are averaged or balanced, it is helpful to remember that:

- ► On a day-to-day basis, buyers strive to achieve a projected markup by the averaging process.
- ► Appropriate cumulative markups result from:

    The effective averaging of many purchases.

    The effects of this averaging on the entire inventory.

- ► Basic markup equations are applied to given information for calculation of the figures required for solutions.
- ► Solutions can be calculated readily when:

    It is understood what information is missing.

    The given information is determined.

    The calculations required for the solution (based on the given information) are identified.

The next section's problems and solutions employ the above principles in two ways; that is, the steps necessary for the calculations are listed in sequence, and the results of these calculations are arranged in a diagram. It should be understood that an average markup is always determined by working with a total cost figure and a total retail figure.

A. Averaging Costs When Retail and MU% are Known

This merchandising technique involves one retail price with two or more costs. It is common to have one retail line that is composed of merchandise with varying wholesale costs. Accordingly, the merchandiser must be able to calculate the proportion of merchandise at different cost amounts that can carry the same retail price, and still achieve the desired markup percentage. To understand this concept fully, it is essential to realize that the merchandiser is attempting to proportion the varying costs to achieve the desired markup percentage, because the aggregate results are the buyer's major concern.

PROBLEM:

For a special sale, a sportswear buyer plans to promote a $25 vest. Consequently, the buyer purchases 500 vests, and wants to achieve a 52% markup. An order for 100 vests that cost $10.75 each is also placed by this same buyer. What will be the average cost of the remaining pieces?

SOLUTION:

Given information is in **bold**, and the other items are calculated from the basic markup formulas; each calculation is identified with the steps of the procedure.

Step 1: Find the total planned retail.

500 pcs. × **$25** = $12,500

Step 2: Find the total cost of total planned retail from planned MU percent.

C. = R. × (100% − MU%)

C. = $12,500 × (100% − **52%**)

C. = $12,500 × 48%

C. = $6,000

Step 3: Find the cost of purchases to date.

100 pcs. × **$10.75** = $1,075

Step 4: Find purchase balance in units and dollars.

Purchase balance = Total planned figures − Purchases to date.

(500 pcs. cost $6,000) − (**100 pcs. cost $1,075**) = (400 pcs. cost $4,925)

Step 5: Find the average cost of purchase balance. Average cost of purchase balance = $ Cost on balance of purchase ÷ Purchase balance number of units $4,925 ÷ 400 pcs. = $12.31

	Total Pcs.	Total $Cost	Total $Retail	MU%
Total plan	**500** (1)	$6,000 (2)	**$12,500** (1)	52%
− Purchases to date	**−100** (3)	**−1,075** (3)		
Purchase balance	400 (4)	$4,925 (4)		

$4,925 ÷ 400 pcs. = $12.31 each is the AVERAGE COST of remaining pieces. (5)

► **For assignments for the previous section, see practice problems 23-27.**

B. Averaging Retail(s) When Costs and MU % are Known

In merchandising, a buyer must be able to manipulate markups because purchases may be made that have two or more costs and the buyer may want to determine an average retail that will achieve the desired markup percent. Furthermore, the buyer may make purchases that have not only two or more costs, but may also have two or more retails. These situations require the proper proportion of varying retails to achieve the planned markup percent because a buyer is always concerned with the aggregate results. The following problems illustrate the averaging processes.

PROBLEM I:

At the end of the season, a swimwear manufacturer offered a retailer merchandise that consisted of 50 two-piece solid-color suits that cost $16.75 each, 75 nylon tank suits that cost $12.75 each, and 40 one-piece skirted suits that cost $14.75 each. In evaluating the value of this offer, the average retail price on each suit was calculated, based on the buyer's goal of obtaining a 49% markup on this purchase. What would be the average retail price of each suit?

SOLUTION:

Given information is in **bold**, and the other items are calculated from the basic markup formulas; each calculation is identified with the steps of this procedure.

Step 1: Find the total cost and total units of purchase.

50 suits @ **$16.75** = $ 837.50
75 suits @ **$12.75** = 956.25
40 suits @ **$14.75** = + 590.00
165 suits = $2,383.75

Step 2: Find the total retail from the planned MU percent.

R. = Cost ÷ (100% – **49%**)
R. = $2,383.75 ÷ 51%
R. = $4,674.02

Step 3: Find average retail for each piece.

Average retail = Total retail ÷ Total units purchased
($4,674.02 ÷ 165 pcs. = $28.33)

	Total Pcs.	Total $Cost	Total $Retail	MU%
Total Plan	165	**$2,383.75**	$4,674.02	**49%**
	(1)	(1)	(2)	

$4,674.02 ÷ 165 = $28.33 retail for each   (3)

PROBLEM II:

The men's apparel buyer bought 50 plaid sportcoats that cost $32.75 each and 80 check sportcoats that cost $42.75 each. An overall markup of 52.5% is needed. If the buyer retails the plaid sportcoats at $65 each, what must be the average retail price for each check sportcoat to achieve the planned markup percent?

SOLUTION:
(Given information is in **bold**, and the other items are calculated from the basic markup formulas; each calculation is identified with the steps of the procedure.)

Step 1: Find the total cost and total units.

50 plaid sportcoats @	**$32.75**	=	$1,637.50
80 check sportcoats @	**$42.75**	=	$3,420.00
130 plaid & check sportcoats		=	$5,057.50

Step 2: Find the total retail from planned MU percent.

R. = Cost ÷ (100% – **52.5%**)

R. = $ 5,057.50 ÷ (100% – **52.5%**)

R. = $ 5,057.50 ÷ 47.5%

R. = $10,647.37

Step 3: Find the total on established retail.

(50 plaid sportcoats × $65 = $3,250)

Step 4: Find the purchase balance in units and dollars. Purchase balance = Total planned figures (minus) – Total established retail figures

130 pcs.	@	$10,647.37
– 50 pcs.	@	$ 3,250.00
80 pcs.	@	$ 7,397.37

Step 5: Find the average retail of purchase balance. Average retail of purchase balance = Purchase balance $ retail ÷ Purchase balance number of units

($7,397.37 ÷ 80 pcs. = $92.47)

	Total Pcs.	Total $Cost	Total $Retail	MU%
Total plan	**130** (1)	**$5,057.50** (1)	$10,647.37 (2)	**52.5%**
– Established retail	–50 (3)		– 3,250.00 (3)	
Purchase balance	80 (4)		$ 7,397.37 (4)	

$7,397.37 ÷ 80 pcs. = $92.47 each for av. retail of remaining pieces.   (5)

▶ **For assignments for the previous section, see practice problems 28-29.**

C. Averaging MU% When Retail and Planned MU% are Known

Realistically, a profit goal can be realized only when a buyer is able to determine the markup percent that must be obtained on present or future purchases to balance out the markup percent that has already been achieved on past purchases. This involves the proportioning of markup percentages on goods so that the aggregate results produce the desired, planned markup percent.

PROBLEM:

A coat buyer plans to buy $5,000 (retail) worth of merchandise. This same buyer requires a 49% MU. The first purchase is 50 coats costing $31.75, with a planned retail of $55. What MU percent should the buyer obtain on the balance of the purchases to attain the planned MU goal of 49%?

SOLUTION:

Given information is in **bold**, and the other items are calculated from the basic markup formulas; each calculation is identified with the steps of the procedure.

Step 1: Find the total cost of total planned retail from planned MU%.

C. = R. × (100% − MU%)

C. = **$5,000** × (100% − **49%**)

C. = **$5,000** × 51%

C. = $2,550

Step 2: Find the total cost and retail of purchases to date.

50 suits @ **$31.75** = $1,587.50 Total cost

50 suits @ **$55** = $2,750 Total retail

Step 3: Find purchase balance at cost and at retail.
Purchase balance = Total planned $ cost and $ retail amounts − The respective figures of purchases to date.

$2,550 − $1,587.50 = $962.50

$5,000 − $2,750 = $2,250

Step 4: Find MU % on purchase balance.

$ MU = R. − C.

$ MU = $2,250 − $962.50

$ MU = $1,287.50

MU % = $ MU ÷ R.

MU % = $1,287.50 ÷ $2,250 = 57.2%

	Total $Cost	Total $Retail	MU%
Total plan	$2,550 (1)	**$5,000**	**49%**
− **Purchases to date**	−1,587.50 (2)	−2,750.00 (2)	
Purchase balance	$ 962.50 (3)	$2,250.00 (3)	

$1,287.50 ÷ $2,250 = 57.2% on purchase balance   (4)

▶ **For assignments for the previous section, see practice problems 30-34.**

NOTES

23. A glove buyer plans to purchase 120 dozen pairs of gloves for a pre-Christmas sale. The unit retail price is planned at $17.50 and the markup goal for the purchase is 48%. 40 dozen pairs are purchased at the Slimline Company showroom, at $110 per dozen.

    (a.) What is the most the buyer can pay for the balance of the total purchase?

    (b.) What will be the average cost per dozen on the gloves (80 dozen) yet to be purchased?

24. A sportswear buyer — who operates on a 51% markup — needs 300 skirts to retail at $32 each and 180 sweater vests to retail at $25 each. If this buyer pays $11.75 for each sweater vest, how much can be spent for each skirt, without deviating from the target markup percentage?

25. A buyer who needs $10,000 worth of merchandise at retail for a housewares department has written orders for $2,875.50 at cost. The planned departmental MU percentage is 53.5%. How much (in dollars) is left to spend at cost?

26. A December promotion of 1,500 sweaters to retail at $40 each is planned. The buyer requires a 48.5% average markup and has made an initial purchase consisting of 1,200 units costing $23 each. What is the cost to be paid on each remaining unit? Comment on the buyer's "predicament" if you detect one.

27. A buyer for the exclusive Britique Shop purchased 100 cashmere pullover sweaters at $41 each and priced them at $90 retail. Also planned is a purchase of 72 shetland/mohair blend bulk knit pullovers to retail at $50 each. Departmental goal markup is 51.8%. How much can be paid for each shetland/mohair pullover?

28. A buyer purchases a job lot of 400 pairs of men's jeans, 280 pairs costing $14 each, and 120 pairs costing $9 each. If a 51% markup is targeted, what would be the average unit retail price on the lot?

29. A buyer purchased 120 maillot swim suits at $32 cost and placed a $65 retail on them. 60 string bikinis are also purchased at $14 each. What would be the retail price on the bikinis if a 50% markup is desired on the combined purchase?

30. A buyer plans to purchase coats at a 48% markup with a retail value of $18,500. If the buyer acquires 100 coats at $69.75 each and retails them at $125 each, what markup percent must now be obtained on the balance of the purchases to achieve the desired markup?

31. A buyer, who needs $23,000 worth of goods at retail for May, has a planned markup of 49.5%. On May 10, the orders to date total $5,500 at cost and $9,800 at retail. What markup must now be obtained on the balance of the purchases for May to achieve the planned markup for the month?

32. The men's outerwear buyer, who needs a 51% average markup for this department, is planning to buy $9,500 worth of merchandise at retail for the month of January. To date, 100 imported raincoats costing $21.75 each have been purchased, with a plan to retail them at 48% markup. What markup percent is needed on the balance of the purchase to attain the average markup percent?

33. A suit buyer who plans sales of $75,000 at retail during April, has an average markup goal of 49%. An order is placed with the B&C Sportswear Company for April delivery in the amount of $5,975 at cost and $11,000 at retail. What markup must be made on the balance of the April purchases to achieve the planned markup?

34. A housewares buyer plans a $15,000 promotion of decorative stepstools to retail at $25 each at a 48% markup. A local manufacturer supplies 500 stools for $14 each. What markup percentage must be obtained on the remainder of the planned purchase to reach the markup goal?

► III. LIMITATIONS OF THE MARKUP PERCENTAGE AS A GUIDE TO PROFITS

From the discussions on markup and its relationship to profit, it is understood that markup can be calculated in both dollars and percentages. A markup percent is a useful guideline in establishing a markup high enough to cover expenses, provide for reductions, and still realize a profit.

In the calculation of all types of markup, the focal point is generally on the markup percent because it is so meaningful for analysis and comparison. However, under certain conditions, attention must be given to the dollar markup. Expenses are paid in dollars and profit is invested, reinvested, or taken to the bank in dollars. During periods of declining sales, a markup percent initially deemed appropriate, may be insufficient for a profit because a certain markup percent is based on a specific estimated sales figure. There is a correlation between these two figures. Conversely, as sales increase, the markup percent can decline even though the dollar markup is maintained at an adequate level.

For example, a buyer may attain an initial markup of 52% on merchandise sold during a period, such as $300,000 at retail, $144,000 at cost. The markup of $156,000 is large enough to cover expenses of $126,000 and markdowns of $12,000 and allow a profit of $8,000.

Initial markup		$156,000 ($300,000 × 52%)
− Expenses	−136,000	
+ Reductions	+ 12,000	−148,000
Total profit		$ 8,000

If sales declined 10% to $270,000, the 52% markup may not yield a profit. As sales decline, the fixed nature of expenses will not decline proportionately to the sales, if at all, and it is not always possible to lower markdown reductions, which may increase to stimulate lagging sales. This type of situation can result in a loss despite the establishment of what seemed a satisfactory markup percent.

The results are illustrated in the following example:

With the same 52% markup, only $140,400 would be available to cover the expenses of the $136,000 and reductions of $12,000.

Initial markup		$140,400 ($270,000 × 52%)
− Expenses	−136,000	
+ Reductions	+ 12,000	−148,000
		−$ 7,600 (Loss)

Therefore, an astute retailer recognizes that the dollar markup figure can not be ignored while attempting to achieve a fixed markup percent.

NOTES

35. Determine the initial markup for a designer boutique that has the following projections for the next season:

Planned sales	=	$4,000,000
Markdowns	=	600,000
Expenses	=	38%
Cash discounts	=	4%
Shortages	=	20,000
Profit	=	5%

36. What is the cumulative markup to date for an accessories department whose figures are:

	Cost	Retail
Opening inventory	$360,000	$780,000
Purchases resource X	60,700	120,500
Purchases resource Y	120,000	245,000
Purchases resource Z	150,000	325,000

37. A shoe department with an initial markup of 53.2% had total retail reductions of 12%. What maintained markup was achieved in this department?

38. A men's outerwear department, with an initial markup of 54%, needs to maintain a markup of 48%. How much can be allowed for reductions?

39. In problem #38, determine if the 48% markup would be obtained based on the reduction percentage that was calculated.

40. The owner of a children's store wishes to increase annual profits (which were 2.5% of net sales) from $5,000 to $15,000. The operating expenses are anticipated to increase from 30% to 32% because of increased promotional events. The total retail reductions are planned at 15.5%, the same as last year, and there are no cash discounts or alteration costs.

(a.) What was last year's initial markup?

(b.) To achieve the owner's increased profit on the same sales volume, what initial markup should be planned for this year?

41. A buyer wishes to buy 3,500 novelty scarves for a special import promotion. The item is to retail at $16.95 each with a planned markup of 54%. If the cost for each of 2,100 scarves is $8.75, how much may be paid for each scarf not yet purchased?

42. A luggage buyer plans a European buying trip to purchase $250,000 (cost value) of assorted travel packs. On the first stop in London, orders are placed for $92,000 (at cost) for an assortment of leather luggage, which will be priced at $180,000 retail. If the average markup is targeted at 52%, what markup percent should now be taken on the balance of the European orders?

43. A men's furnishings buyer purchases a closeout of 1,500 velour shirts offered at $9.85 each and 950 knit shirts at $5.80 each. The planned markup on the purchase is 49.5%. If a unit retail of $17.95 is placed on the velour shirts, what is the lowest possible retail price that the knit shirts may be marked, to remain within the markup framework indicated?

44. A buyer, who has a departmental markup of 49%, wants $40,000 (retail) worth of men's casual shirts for next season. An order for turtlenecks is placed in the amount of $6,400 at cost and $11,000 at retail. What markup percent must now be achieved on the balance of the purchases?

45. A lingerie department made a purchase of 100 dozen nylon briefs costing $36/dozen, 75 dozen stretch bras costing $54/dozen and 12 dozen teddies at $84/dozen. If the briefs are retailed at $6 each, and the bras at $9 each, at what retail price should the teddies be marked to attain a 53% markup?

46. A piece goods buyer, who needs a markup of 50%, plans to purchase $5,000 worth of goods (retail value) during the month of November. At the middle of the month, the total of the invoices for the purchases to date (at both cost and retail) for the month to date amount to $2,000 at cost and $3,700 at retail. What markup in dollars and percent must be obtained on the balance of the purchases?

47. A buyer needs 5,000 umbrellas for the 37th annual storewide "April Showers" promotional event, to retail at $14 each. The planned markup is 49%. If 3,500 units are bought at one resource for $7.70 each, how much can the buyer afford to pay for each of the remaining 1,500 units?

48. A children's apparel buyer is offered a lot of coats consisting of 125 toddler boys' coats at $20 each and 160 toddler girls' coats at $23 each. All the coats will be sold at the same retail price. If the buyer is to make a 54% markup on the entire transaction, at what price must each coat be retailed?

49. A men's accessories buyer purchases 10,000 neckties from a manufacturer — who is relocating — at a cost price of $10 per tie. Noticing that the ties are of two types, 6,200 solids and 3,800 stripes, the buyer decides to attempt some creative merchandising. If each solid tie is retailed at $18, what retail price should be placed on each striped tie if an average markup of 51% is desired?

50. An outerwear buyer confirms an order reading as follows:

    ▶ 145 tailored raincoats costing $47 each.

    ▶ 75 three-quarter-length rainjackets costing $26 each.

If a retail price of $85 is placed on the tailored coats, and a markup average of 48% is sought, what retail price must the rainjackets carry?

51. The boyswear buyer purchases 20 dozen pairs of jeans at $108/dozen and 15 dozen flannel shirts at $96/dozen. The departmental markup is planned at 52%. If the jeans are priced at $18, what price must be put on the shirts to reach goal markup?

52. A buyer purchases a closeout of 1,000 coats at a cost of $46 each, which are divided into two groups: 425 belted coats and 575 unbelted. If the desired average markup is 49%, what retail price should be placed on each of the unbelted coats if the belted ones are priced at $95 each?

53. A junior sportswear department shows an opening stock figure of $540,000 at retail, owned at a 48% markup. To date, since the last stock determination, new purchases were $1,700,000 at retail, with a 50% markup. Find the cumulative markup percent on total merchandise handled in this department to date.

54. The gross margin in a handbag department is 42.7% (40% expenses + 2.7% profit), which shows reductions (total of markdowns, shortages, and employee discounts) of 14.5%. State the initial markup percentage.

55. The planned gross margin in a young men's sportswear department is 46%, with reductions of 15%, cash discounts earned equalling 6%, and alteration costs amounting to 2%. What is the initial markup percentage?

56. A men's apparel department planned an initial markup of 49% and reductions of 16%. Find the maintained markup.

57. For a Columbus Day event, a buyer plans to purchase 250 knit dresses to retail for $75 each. A 47% markup is needed on the total purchase. From one of the buyer's best resources, 110 dresses are purchased that cost $33.75 each, even though this markup percentage is less than what ultimately must be achieved. The buyer feels, however, that the costs will be averaged on the balance of other purchases from alternate resources.

(a.) What must be the average cost per dress on the balance of the purchases if the buyer is to attain the planned MU percentage?

(b.) What MU percentage did the buyer attain on the first purchase of 110 dresses?

(c.) What MU percentage did the buyer have to get on the balance of the purchases to realize the overall needed MU?

58. For the month of July, a women's sportswear department had net sales of $250,000. The billed cost of the goods, including transportation, was $120,000. Cash discounts amounted to $9,600 and there were alteration costs of $400. Calculate (in dollars):

(a.) The maintained markup.

(b.) The gross margin.

Targeting Gross Profit

Miss Jay Tee is the shoe buyer for a fashion specialty store located in the trendy "Near-North Side" of Chicago. It is October 1 and she is scheduled for a European buying trip on October 25, during which she will place orders for Spring delivery. Her Spring six-month plans are complete and have been approved by management.

Before leaving on her buying trip, she reviews her current operational figures to evaluate her position for the forthcoming last quarter of the year (i.e., November, December, January). They are:

- ► Merchandise on hand (BOM) November 1 — $1,300,000 with a 51.5% cumulative markup.
- ► Closing inventory (BOM) February 1 — $400,000.
- ► Planned sales:

November	December	January	Total Sales
$500,000	$550,000	$250,000	$1,300,000

- ► Planned markdowns to be taken in November, December, January — $80,000 or 6.2%.
- ► Estimated shortages are 2.0% for this period.
- ► Gross margin (no cash discounts or alteration costs) — 48.1%.

Based on this information, Miss Tee did some calculations and determined that she still had $506,000, at retail, to spend for this quarter.

It is customary for Miss Tee to devote several days at the beginning of her trip to scout the market to obtain extremely desirable current seasonal goods, available for immediate shipment, at advantageous prices that will stimulate pre-Christmas sales and furthermore, will generate sales in the slow-selling month of January, traditionally associated with only clearance and/or highly competitively priced merchandise.

On October 25, she arrives in Florence, Italy, and as she covers the boot resources, she finds there is an abundance of the current season's merchandise — available for immediate shipment — due to cancellations by importers and retailers. This situation has been caused by sluggish economic and business conditions. One of her best key resources has in stock 4,000 pairs of leather boots in this season's avant garde styling with balanced sizes and colors. 50% of the resource's offering are the same styles as 500 pairs currently in stock, selling reasonably well at $200 retail. Providing she agrees to buy the complete group, the final landed cost negotiated is $50 per pair. Despite the fact that she considers this merchandise to be superb value at this sensational cost, she is hesitant about buying such a large quantity so late in the selling season. However, she receives a fax from her assistant informing her that the boots are moving quickly and sales have increased due to very cold, wintery weather conditions. She decides to buy this offering before a competitor has an opportunity to do so. Her strategy considerations are:

- ► Retailing the group.
- ► Shipping for present selling.
- ► Integrating this purchase with present stock.
- ► Obtaining the necessary markup percentage.

She commits herself to the purchase, returns to her hotel and diligently works to map out the delivery and pricing factors. Her first decision is to start shipping some of these goods so they are available, as soon as possible, for November selling. Therefore, despite the expense, she immediately sends 1000 pairs by air because it is fast and it serves her purpose. This mode of transportation increases the delivered cost of the 1000 pairs from $50 to $55 a pair, and the retail is set at $150.

Having made this decision, she turns her attention to the 500 pairs in stock currently selling at $200 a pair. Because markdown money is available, Miss Tee wants to begin to reduce the higher-priced boots while there is traffic. She can afford to reduce these 500 pairs from $200 to $150, taking a larger than usual first markdown, because the markup of the 1000 pairs (that will sell simultaneously) is so above average.

On the balance of the 3000 pairs, she divides half for December selling and retails them for $125. The remaining 1500 pairs she reserves for her annual Blizzard Promotion scheduled in January. For this event, her boot sales rely solely on merchandise that create "riots". She did not yet set a retail price on the 1500 pairs for the January sale because she would like to price them at levels that are somewhat low, but will not kill the required markup for the season. However, she calculated that a 54.4% markup was needed on the entire purchase so that her seasonal markup target will be met.

Prepare a merchandising plan for class discussion and in doing so, consider the following:

▶ Do you agree with the merchandising techniques Miss Tee used in this particular purchase? If so, why? If not, why? Justify your position mathematically.

▶ Advise Miss Tee of the retail price you would establish on the 1500 pairs of boots targeted for the January promotion, keeping in mind that all of her objectives should be satisfied.

UNIT IV
The Retail Method of Inventory

▶ Knowledge and understanding of the retail method of inventory.

▶ Differentiation between physical inventory and book inventory.

▶ Identification and recognition of procedures necessary to implement the retail method of inventory.

▶ Calculation of book inventory figures at cost.

▶ Identification and description of the forms used in the retail method of inventory.

▶ Recognition and identification of the causes of overages and shortages.

▶ Calculation of overages and shortages based on inventory figures.

▶ Evaluation of the advantages and limitations of the retail method of inventory.

KEY TERMS

book inventory

closing book inventory

debit memo form

journal/purchase record

perpetual inventory

physical inventory

opening book inventory

overage

price change form

retail method of inventory

running book inventory

shortage

transfers

To control and guide the operations of a retail organization it is essential to keep records. Records are the working tools that provide information on the profitability of a business or for making everyday decisions (e.g., what types of merchandise are needed, when and how much is needed, etc.). Of particular concern to management and buyers is inventory control.

Successful merchandising requires that the size of stocks offered the consumer be large enough to satisfy demand, but that the dollar investment be kept as low as possible. This can be accomplished only by having a frequent indication of stock-on-hand. The retail merchandiser who is concerned with the question "How much can I sell?" must know how much to buy to maintain this satisfactory relationship between the amount of sales volume and the size of stocks carried.

In large retail stores, it would be inconvenient and prohibitive in terms of cost to constantly determine the value of the amount of stock-on-hand by taking an actual count. However, because this balance of sales to stocks is vital, a system of accounting that determines the probable amount of stock-on-hand at any given time — without physically counting the goods — has been devised. This retail system of accounting is called the RETAIL METHOD OF INVENTORY.

The retail method of inventory is a method of averages, with the retail stock figure at the end of an accounting period providing the basis for determining the cost value of stock. The conversion of the closing inventory at retail to a cost figure is calculated by the determination of the cumulative markup percent on the total merchandise handled as described in Unit III. To understand the retail method of inventory, it is vital to realize that it operates on the theory that the merchandise in stock is always representative of the total merchandise handled to date (i.e., stock plus new purchases). It allows an acceptable cost value of the book inventory to be established so that gross margin can be determined periodically.

It is common practice for large stores to "think at retail" because primarily, net sales (100%) are the basis for the analysis of all the relationships of expenses to sales, and are the ultimate determining factors that show whether the merchandising endeavors result in a profit or a loss. Gross margin, which is the difference between the cost of goods sold and the income received from this merchandise (i.e., net sales) is also expressed as a percentage of net sales. Consequently, the danger of failing to make the correct percentage comparisons is eliminated because these percentages are all calculated on the same base (i.e., net sales), which is a retail figure. The dollar value of the inventory owned must also be expressed as a retail figure to predetermine the desired relationship of these two factors (i.e., sales to stocks. See Unit I.) Secondly, in the process of buying and selling to yield a more satisfactory profit, retailers can contrast and compare their merchandising operations with that of other retailers. They compare such factors as net sales produced (a retail figure); the relationship of retail stock to net sales (figures used to attain the needed proportion); the pricing of merchandise expressed as a percentage based on the retail, and the percentage of reductions (figures needed to revise prices originally set on merchandise). Finally, when filing income taxes, insurance claims, etc., the current retail price of the merchandise is the significant valuation.

These three examples illustrate the importance of the maintenance of a perpetual retail inventory figure. Though inventory figures are not always perpetually derived, they can be obtained as often as is desirable, usually every week or month.

I. AN EXPLANATION OF THE RETAIL METHOD OF INVENTORY

The retail system of merchandise accounting permits the retailer to determine the value (at retail) of the stock-on-hand at frequent and periodic intervals without taking constant physical counts. However, it must be noted that periodic — generally semi-annual — physical counts (i.e., inventories) are taken at the current retail prices of the merchandise on hand. To control the stocks and to determine the profitability of individual departments, the retail method of accounting is applied separately for each department. The retail method of inventory valuation involves:

▶ Taking a physical inventory count to determine the total retail value of a particular department.

▶ Determining the cumulative markup percentage on the total merchandise handled.

▶ Deriving the cost value of the closing inventory from the retail by using the cumulative markup percent achieved on the total merchandise handled. Subsequently, this valuation is used to find the cost of goods sold to establish the gross margin.

This system requires the collection and analysis of data pertaining to any movement of merchandise from the time it is bought until it is sold to the consumer. The retail method of inventory requires maintaining a book inventory at retail, as well as other records that permit the calculation of the cost of total merchandise handled during the period. This, in turn, allows the constant calculation of the gross margin amount including the possible protection of profitability. All additions to and deductions from stock must be recorded in dollar values. The computation from "statistical records" or book figures of the amount of merchandise that should be on hand, at retail, is called a BOOK INVENTORY (STATISTICAL INVENTORY). At the beginning of the accounting period under consideration, a PHYSICAL INVENTORY COUNT is taken at the current retail price of the goods owned. It is common that when large stores take a semi-annual physical count of stocks and record the value at retail, that date of receipt of merchandise into stock is also recorded. For example, Figure 12 illustrates the information generally recorded during a physical inventory count.

When taking a physical inventory under this method, it is not necessary to list the cost price of each individual item. The physical count of each individual item, at the retail price stated on its ticket, is recorded and the total retail figure (e.g., $100,000) is the actual amount of goods accounted for at the time the semi-annual count is made. If warranted, this count can be made more often. This actual physical count figure is then used as the closing physical stock figure for that accounting period. If the physical inventory value is less than the book inventory, the difference is called a SHORTAGE. When the physical inventory value exceeds the book inventory, that difference is called an OVERAGE.

II. GENERAL PROCEDURES FOR IMPLEMENTING THE RETAIL METHOD OF INVENTORY

A. Finding an Opening Inventory Figure

When the "retail method" system is first installed, a complete physical count, at retail (see Figure 9), is taken for each merchandising department. (Please

FIXTURE: _____

00/00/00

INVENTORY : 012
LOCATION : 0001
SEQ : 000836
CHECK DIGIT : 0

012000010008360

INITIAL COUNT

| QTY | EXEC-CHECK |
| BY: _____ | BY: _____ |

RECHECK COUNT

| QTY | EXEC-CHECK |
| BY: _____ | BY: _____ |

SHEET : 0 1 2 0 0 0 1 0 0 0 8 3 6   CHECK DIGIT : 0

LN	DEPT	CL	UPC / SHORT–SKU	AGE	RETAIL	QTY	DESCRIPTION	DEDUCT EX CK
01					.			01
02					.			02
03					.			03
04					.			04
05					.			05
06					.			06
07					.			07
08					.			08
09					.			09
10					.			10
10					.			10
11					.			11
13					.			13
14					.			14
15					.			15
16					.			16
17					.			17

012   0001   000836

FIGURE 9 Physical Inventory Count Sheet

note that this physical count reflects the retail value of the inventory at aggregate retail prices.) The retail value of the goods counted is used as the OPENING BOOK INVENTORY figure. This is the same figure as the CLOSING PHYSICAL STOCK figure of the previous accounting period. Inventory counts continue to be taken semi-annually at the current retail prices of the goods. (Remember, closing physical inventory equals the opening book inventory for the next period.)

B. Maintaining a Perpetual Book Inventory Figure

A book inventory indicates the amount of stock in dollars that has been determined from records rather than from an actual count. During the time period between the semi-annual physical counts just described, many other merchandising transactions occur, such as:

- ▶ Goods are sold to customers.
- ▶ Merchandise is purchased from resources and received.
- ▶ Customers return merchandise.
- ▶ Merchandise is marked down.
- ▶ Goods are transferred to other departments in the same store for resale.
- ▶ Merchandise is transferred to other stores.
- ▶ Merchandise is returned to vendors occasionally.

Every such transaction, and any other kind of merchandise movement is accompanied by paper work in the form of saleschecks, return-to-vendor forms, orders, etc. The retail organization's statistical department or the accounting division records every transaction and adds to or reduces the "book stock" accordingly. For example, sales to customers or transfers to other departments reduce the "book stock" (also known as PERPETUAL or RUNNING BOOK INVENTORY) by the dollar amount corresponding to the retail value of the goods sold or transferred. The records must be current and accurate so that a retail book inventory figure is always available during an accounting period. Any additions to or deductions from stock must be recorded at the current retail prices and reported to the statistical department. In other words, every change that affects the stock value must be recorded.

In a multi-store operation, a constant inventory figure is maintained for the particular department in each store. This inventory figure is evaluated, not only on an individual store basis, but also on a department's overall operation.

The following is an example of a department for a branch store, which should serve to illustrate the "addition or deduction" of the value of the merchandise in stock. All these changes must be recorded to ensure accuracy.

EXAMPLE:

Increase of Retail Value		Decrease of Retail Value	
Purchases (total retail value of mdse. received)	$35,000	Net sales	$28,000
+ Transfers in	+ 5,000	+ Transfers out	2,000
Total stock additions	$40,000	+ Ret. to vendor	3,000
(Total ins)		+ Empl. disc.	500
		+ MD differences[1]	+ 500
		Total stock deductions	$34,000
		(Total outs)	

---

[1]Markdown differences are included in stock deductions because they lower the value of the inventory.

1. Calculating a Book Inventory Figure at Retail

A retail book inventory is determined by:

▶ Taking a periodic physical count, at retail, to determine a closing physical stock figure.

▶ Establishing the physical count amount as the opening retail book inventory.

▶ Adding all new purchases at retail to determine the total merchandise handled at retail.

▶ Subtracting all retail deductions (e.g., net sales, markdowns, etc.) from the retail total merchandise handled to find the closing book inventory at retail.

CONCEPT:

Book Inventory at retail = Add the opening physical inventory figure to the net retail purchases (Total Merchandise Handled) and any other stock additions; then, (from the resultant sum) subtract net sales, markdown differences, and any other deductions from stock.

PROBLEM:

On January 31, the physical count of the infants' department in Branch #15 revealed an inventory of $20,000. On February 1, the opening retail inventory of this department was $20,000. From February 1 to July 30, retail purchases amounting to $40,000 were received. The net sales for this period were $30,500, the markdowns taken were $2,300, employee discounts were $200, returns to vendors were $350, and transfers out were $750. What was the retail book inventory for this period under consideration?

SOLUTION:

Opening retail inventory (Feb. 1)		$20,000
+ Retail purchases (Feb.1 - July 31)		+40,000
Total merchandise handled (ins)		$60,000
Net sales	$30,500	
+ Markdown differences	+ 2,300	
+ Employee discounts	+ 200	
+ Returns to vendor	+ 350	
+ Transfers out	+ 750	
Total deductions (outs)	$34,100	–34,100
Retail Book Inventory	=	$25,900

2. Calculating a Book Inventory at Cost[2]

A book inventory, at cost, is determined by:

▶ Converting the opening retail book inventory to a cost inventory figure by using the cumulative markup percent achieved on the stock plus purchases during the previous accounting period.

---

[2]The retail method of inventory assumes that the average markup on the closing inventory is the same as the markup on the total merchandise handled.

▶ Adding all new purchases, at cost[3], plus freight (on cost only) to the cost opening inventory figure to determine the total merchandise handled at cost.

▶ Converting the closing book inventory at retail to a closing book inventory at cost, by using the cumulative markup percent that has been calculated from the difference between the total merchandise handled at retail and total merchandise handled at cost.

PROBLEM:

Calculate a book inventory at cost, if the markup percentage on the previous season's inventory is 51% and the cost of the new purchases is $19,800. (The same figures are used as in the previous problem illustrating the determination of a book inventory figure at retail.)

SOLUTION:

1. Convert opening retail inventory figure to cost inventory figure.

$$\text{Opening cost inventory} = \$20,000 \text{ R} \times (100\% - \text{MU}\%)$$
$$= \$20,000 \times (100\% - 51\%)$$
$$= \$20,000 \times 49\%$$
$$= \$ \ 9,800$$

2. Add all new purchases plus freight to opening inventory at cost.

Given:

New purchases	=	$19,800
+ Freight	= +	200
		$20,000

Opening inventory at cost	=	9,800
+ New purchases (including freight) at cost	=	+20,000
Total merchandise handled at cost	=	$29,800

3. Find cumulative markup percentage on total merchandise handled.

	Cost	Retail
Opening inventory	$ 9,800	$20,000
+ Purchases	+ 20,000	+40,000
Total merchandise handled	$29,800	$60,000
Markup	= $60,000 Retail	
	− 29,800 Cost	
	$30,200 MU	$30,200 MU = 50.3%
		$60,000 Retail

4. Convert closing book inventory at retail to closing book inventory at cost by using the cumulative MU percentage.

$25,900 Closing book inventory at retail × (100% − 50.3%)
$$= \$25,900 \times (100\% - 50.3\%)$$
$$= \$25,900 \times 49.7\%$$

Closing Inventory at Cost $$= \$12,872.30$$

▶ **For assignments for the previous section, see practice problems 1-11.**

---

[3]Generally, purchases are accumulated over a specific period of time, (e.g., a six-month season) not month-by-month, and then are added to the opening inventory at cost to find the total merchandise handled at cost.

C. Forms Used in the Retail Method of Inventory

Today a retailer's records are often supplemented and/or obtained through the store's (or organization's) computer system, (e.g., POS markdowns). How the information is recorded is a matter of choice or monetary necessity, but the loss of records or the failure to record the proper information that shows the increase or decrease in the value of the stock will result in an inaccurate stock valuation of the book inventory. There are many forms to help the retailer accurately record this valuable information. Each form serves a particular function, but not all retailers use identical forms. The forms in this section illustrate the function of each record required to appreciate or depreciate the value of a stock. Today, with the widespread use of automated merchandising systems, the information necessary to maintain an accurate recording of transactions is adopted and tailored to each retailer's needs.

FIGURE 10 Journal or Purchase Record

CLS	VND	STYL	RETAIL	TOTAL STORE	01	02	03	04	05	06	07	08	09	10	11	12	13	14	15	16	17	ERR
\multicolumn{24}{c}{**Merchandise Receipts**}
120	0515	0078	52.00	10													10					
120	0525	8514	50.00	252	36	60	60	24	24	24	24											
120	0550	8329	50.00	−1							−1											
120	0550	8428	68.00	20												20						
120	0550	9304	48.00	−1							−1											
120	0700	9005	96.00	14	5	5	4															
120	0700	9009	120.00	18	6	6	6															
120	0700	9021	48.00	72	24	24	24															
120	0717	5011	150.00	30		6	6	6		6								6				
120	0717	5015	110.00	36		6	6			6		6	6					6				
120	0717	5047	30.00	66	6	6	6	6		6		6	6			6		6	6	6		
120	0802	4723	42.00	288	36	40	40	16	16	16	28			16	16	16		16		16	16	
120	0808	3452	48.00	86								10	10	8	8	10	6	10	8	8	8	
120	0808	3454	50.00	90	8		8	6	6		6		8	6	6	7	4	7	6	6	6	
DOLLAR CLASS TOTALS $					6244		9076		2172		2578		1720		1356		1008		864		1356	
			$	52506		8772		3525		3612		1320		1356		3042		3242		1536		
130	0833	8864	14.00	939	96	120	180	72	96	60	108			24	24	60	24		26	24	25	
DOLLAR CLASS TOTALS $					1344		2520		1344		1512				336		336		364		350	
			$	13146		1680		1008		840				336		840				336		
220	0400	3401	115.00	42	10	6	10						8					8				
220	0717	5045	98.00	42	6	6	6	6		6			6					6				
DOLLAR CLASS TOTALS $					1738		1738						1508									
			$	8946		1278		588		588								1508				
DOLLAR DEPT. TOTALS $					9326		13334		3516		4090		3228		1692		1344		1228		1706	
			$	74598		11730		4848		5040		1320		1692		3882		4750		1872	
				STORE	01	02	03	04	05	06	07	08	09	10	11	12	13	14	15	16	17	ERR

FIGURE 11 Outstanding Transfer List Form

```
                              OUTSTANDING TRANSFER LIST AS OF  OCT 08,YY      DATE  OCT 09,YY          PAGE: 1
                                                                             TIME  5:31 PM

1) BY FROM STORE - WITH DETAILS
FROM    TO   TRANSFER      TRANSFER    SKU      STYLE    COL   SIZE   QUANTITY    PRICE     RETAIL      COST     TRANSFER
STORE  STORE NUMBER TYPE   DATE        NO.      NO.                                         AMOUNT      AMOUNT   REASON
----------------------------------------------------------------------------------------------------------------------
 001    002  000002  1     OCT 08,YY   0015149  SW10     030         XS        4    22.55     90.20      40.00    001 Slow Moving
                                       0015156  SW10     030          S        4    22.55     90.20      40.00
                                       0015164  SW10     030          M        5    22.55    112.75      50.00
                                       0015172  SW10     030          L        3    22.55     67.65      30.00
                                       0015180  SW10     030         XL        2    22.55     45.10      20.00

 003    001  000001  1     OCT 07,YY   0014555  200XT    020   R - 40         3   440.00  1,320.00     825.00    003 Fast Moving
                                       0014563  200XT    020   R - 42         3   440.00  1,320.00     825.00
                                       0014571  200XT    020   R - 44         3   440.00  1,320.00     825.00
                                       0014589  200XT    020   R - 46         3   440.00  1,320.00     825.00
                                       0014662  200BK    020   T - 38         1   440.00    440.00     275.00
                                       0014670  200BK    020   T - 40         1   440.00    440.00     275.00
                                       0014688  200BK    020   T - 42         1   440.00    440.00     275.00
                                       0014696  200BK    020   T - 44         1   440.00    440.00     275.00
                                       0014704  200BK    020   T - 46         1   440.00    440.00     275.00
```

© 1993 Richter Management Services, Inc.

1. Journal or Purchase Record

Figure 13 is a JOURNAL or PURCHASE RECORD, which provides a record of the billed or invoiced costs, transportation charges, cash discounts, retail amounts, and the percents of markup for each individual purchase. The names of the vendors, dates of invoices, and the invoice numbers are also entered. Each department checks this record periodically to ensure that the department is being charged or credited with merchandise either entering or leaving a department, and that these amounts are only intended for that department. This report also shows the department classification, vendor, style, and price receipts in units for the total store as well as individual branches. The negative units represent returns to vendors or corrections. The purpose of the report is to allow buyers to examine receipts entered into the computer against copies of the purchase order. Discrepancies are then reported so that appropriate adjustments can be made.

2. Transfer of Goods

A transfer of merchandise involves the movement of goods. When the merchandise leaves a department, the transfer is out; conversely, when the merchandise is received, the transfer is in. When merchandise is transferred from one store to another (e.g., a branch), an OUTSTANDING TRANSFER LIST FORM is used to record the number of units transferred, unit cost, total cost, unit retail, and total retail. Merchandise may also be transferred from one department to another. This record is used to indicate the change of ownership of merchandise. Figure 14 (Outstanding Transfer List form) illustrates the detailed information that an automated merchandising system can furnish on interstore transfers, (e.g., reasons for transfers, size, color, etc.).

BUYERS PRICE CHANGE WORKSHEET

SYSTEM GENERATED NUMBER	DEPT.	REGION	TYPE	REASON	EFFECTIVE DATE	END DATE	MANUAL DOCUMENT #

REGION: 01-ALL SELLING LOCATIONS / 02-07.22 / 03-19 ONLY

INFO ONLY – ☐ YES OR ☐ NO   (TYPE FOUR PRICE CHANGE ONLY)   REFERENCE #

SPECIAL INSTRUCTIONS:

NON-ITEM DEPARTMENTS ALL IN SHADED AREA

CHECK OFF STORE CODE IF PRICE CHANGE IS NOT FOR ALL LOCATIONS

01-NY	06-SH	10-PGA	14-TC	18-N/A
02-BR	07-CP	11-CH	15-N/A	19-WH
04-NA	08-NM	12-WP	16-KP	21-FA
05-BC	09-GC	13-WF	17-WG	

LINE #	CLASS #	VENDOR #	MARK STYLE # (NO CHECK DIGIT)	COLOR #	SIZE #	CURRENT OWN RETAIL	NEW OWN RETAIL	ON HAND	ITEM DESCRIPTION
01									
02									
03									
04									
05									
06									
07									
08									
09									
10									

TYPES:
01-MARKDOWN TOTAL STYLE
02-MARKDOWN PARTIAL STYLE
04-PROMOTIONAL MARKDOWN
05-MARKDOWN CANCELLATION
06-MARKUP

REASONS: CLEARANCE
10-FIRST MARKDOWN
11-SECOND MARKDOWN
12-ANY ADDITIONAL MARKDOWN
19-JOB OUT OF STOCK (JOBBER)
20- MARK OUT OF STOCK (SALVAGE)

REASONS: PROMOTIONAL
22-NON POS MD
23-NON POS MDC

REASON CODE 20 (SALVAGE)
WILL NOT UPDATE ITEM OWNED RETAIL, BUT
WILL REDUCE QUANTITY ON HAND AMOUNT

COMPLETE ONLY IF
MD CANCELLATION FOR VENDOR ALLOWANCE

VENDOR HOUSE _____
PRIMARY CLASS _____
CLEARANCE
C.C. NUMBER
C.C. DOCUMENT MUST BE ATTACHED
(NOTE: KEY C C NUMBER IN REFERENCE NUMBER FIELD)

REASON CODE
☐ 21
☐ 13

FIGURE 12 Buyers Price Change Worksheet

FIGURE 13 Computer Price Change Entry

```
DATE: OCT 09,YY
                              PRICE CHANGE ENTRY
=========================================================================

COMPANY  01   GROUP 0001   TYPE 1   FUNCTION 1  DATE 10/09/YY  EFFECTIVE DATE FR: 10/10/YY
                           (New)    (Price Change)             EFFECTIVE DATE TO: 12/31/YY

  1) AUTHORIZATION   :123456
  2) ENTRY TYPE      :1      MARKDOWN
  3) BUYER NO        :001    Michael Wood
  4) REASON          :001    Slow Moving                    :12/31/YY
  5) EFF. DATE FROM  :10/10/YY        6) EFF. DATE TO       :999
  7) REGION FROM     :000             8) REGION TO          :999
  9) STORE FROM      :000            10) STORE TO           :Y
 11) REM–1: Permanent price change   13) PRODUCE TICKETS (Y/N)  :N
 12) REM–2: effective for all stores 14) COUNT INVENTORY

LINE   SEASON  DEPT  CLS  PR  LN  COST  PT  COOR  GR  STYLE  CHANGE      TO RETAIL
                                                            %   AMT     PRICE
001 FROM  L      01                                        10.00
        TO  L      02
EXCLUDE STYLES: 12056       12654        13101        141256
```

© 1993 Richter Management Services, Inc.

3. Price Change Forms

All retail price changes that are required to merchandise a department must be recorded. Among the changes necessary to record are:

- ▶ The number of units.
- ▶ The old retail price per unit.
- ▶ The new retail price per unit.
- ▶ The difference per unit.
- ▶ The total amount of difference.

Today it is common that temporary price changes (e.g., one-day sale items) are recorded by a cash register at the time of the purchase. As described in Unit II, this is called a point-of-sale markdown (POS). When a consumer pays for a purchase, the pre-programmed cash register records the new or lower retail, which corresponds exactly with the prescribed reduction posted on signs displayed with the merchandise. Only permanent reductions are recorded manually on forms illustrated by Figures 12-14. Figure 12 is a worksheet, which requires the listing of any or all price changes. Figure 13 shows the permanent or temporary price change information that has been entered onto a computer screen and Figure 14 illustrates the form that is used by the individual branches.

4. Charge-Back to Vendors

A DEBIT MEMO FORM shown in Figure 15 records the return of merchandise from the retailer to the vendor, which may occur for a variety of reasons. It displays the number of pieces or units, the name of the item, and both cost and retail prices. Typically, a worksheet, as illustrated by Figure 16, is used to record the return to vendor information before it is verified by the person packing the merchandise. This ensures that the actual debit memo is legible and correct.

BRANCH PRICE CHANGE

DOCUMENT NUMBER 12899

REASON: ☐☐   TYPE: ☐☐

STORE NO.: ☐☐☐☐

TOTAL COUNT: ☐☐☐☐

REF-NBR:

EFF-DATE: _____
(DATE KEY ENTERED)

LINE	DPT	CL	VND	MKST (NO CHECK DIGIT)	CC	SIZ	SKU-UPC	CUR-OWN RETAIL	NEW-OWN RETAIL	CNT	DESCRIPTION	PC NUMBER (FROM SCREEN)
01												
02												
03												
04												
05												
06												
07												
08												
09												
10												
11												
12												
13												
14												
15												
16												

REASONS: CLEARANCE
50 - Customer adj
51 - Salvage
49 - Store initiated clearance
42 - Missed markdown

REASONS: PROMOTIONAL
41 - Customer service to meet competition

TYPES:
01 - Markdown
05 - Markdown Cancellation
06 - Markup

(NOTE: ONLY ONE REASON CODE PER PRICE CHANGE PER FORM)

INPUT BY: DATE:

FIGURE 14 Branch Price Change

FIGURE 15 Debit Memo

DEBIT MEMO		
INVOICE		**718584**

POSTMASTER: RETURN REQUESTED

DATE	DEPT.	VENDOR NO.	STORE

CHARGE (PRINT)

STREET (PRINT)

CITY, STATE & ZIP CODE (PRINT)

REFERENCE

A/P AUTH. NO.

SPECIAL INSTRUCTIONS

SEND (PRINT)

STREET (PRINT)

CITY, STATE & ZIP CODE (PRINT)

RETURN ARRANGED WITH	PREPARED BY

OFFICE APPROVAL	TERMS	FREIGHT
	/	VENDOR ☐ DEPT ☐

STYLE	ITEM / SKU #	COLOR	CLASS	QUANTITY	UNIT	COST		RETAIL	
						UNIT	EXTENSION	UNIT	EXTENSION

REASON FOR RETURN AND/OR CHARGE:

1	ADVER-TISING	4	CREDIT AS AGREED	7	NOT AS ORDERED	10	*STOCK REPAIR
2	CANCELLED ORDER	5	DAMAGED	8	NOT ORDERED	11	**CUST. REPAIR
3	COLOR WRONG	6	DEMO SALARY	9	OVER SHIPPED	12	OTHER _____

HANDLING	
SUB-TOTAL COST	
FREIGHT-IN	
FREIGHT-OUT	
TOTAL COST	

TOTAL RETAIL	

* ISSUE NEW INVOICE TO COVER REPLACEMENT OR STOCK REPAIR.
** VALUE OF MDSE. TO BE CHARGED TO YOU IF NOT RETURNED IN 30 DAYS.

SHIP VIA	IF FREIGHT IS PREPAID ENTER THE AMOUNT HERE	$	CARTONS	WEIGHT	PACKED BY	DATE PACKED

NAME OF CARRIER	BILL OF LADING #	NUMBER OF CARTONS	DATE PICKED UP	PICKED UP BY (SIGNATURE)

FREIGHT IN & OUT IS CHARGED TO YOUR ACCOUNT WHEN A SHIPMENT IS MADE CONTRARY TO THE TERMS OF OUR PURCHASE ORDER.

ROUTING INSTRUCTIONS:
WHEN RETURNING MERCHANDISE (UNLESS OTHERWISE SPECIFIED HEREIN), SHIP PER CURRENT ROUTING INSTRUCTIONS.

IF YOU DO NOT HAVE THESE INSTRUCTIONS CONTACT: CORPORATE TRAFFIC OFFICE

WHEN RETURNING MERCHANDISE, ADDRESS SHIPMENT TO THE APPROPRIATE ADDRESS BELOW. BE SURE TO SHOW OUR STORE #, DEPT. # AND INVOICE # **ON ALL CARTONS AND MEDIA.**

FIGURE 16 Return to Vendor Authorization and Worksheet

RETURN TO VENDOR AUTHORIZATION AND WORKSHEET   № 627893

Instructions: (Please print legibly)

Department Manager

DM No. _____.

1) Dept. Mgr. prepares worksheet
2) Complete Worksheet including quantities and extensions
3) Completely prepare D.M. from information
    on Worksheet except for merchandise
    quantities, AND extensions.
    (D.M. Room will do this).
4) Do Not put in cost figures.

Dept._____ Store _____ Prepared by _____ Date _____

Mfr. Name _____ Freight paid by dept. _____ vendor _____
                                                                           CHECK ONE

Vendor No.   _____

Return arranged with _____ Sticker needed   Yes _____ No _____

Special Instructions _____ Terms _____

Reason _____

CHARGE TO:	SEND TO:
Name	Name
Address	Address
City State Zip	City State Zip

Item / SKU No. *	Class	Style	Size	Color	Qty.	Units	Cost Unit	Cost Ext.	Retail Unit	Retail Ext.
Total										

ON DMS OVER $400 RETAIL

COMPLETE THE FOLLOWING:

ACCTS. PAYABLE AUTH. NO. _____

D.M. ROOM COMPLETES

D.M. # _____

Date _____

Total $ Amount _____

Packed by _____

Checked by _____

*IF BSR MERCHANDISE IS INVOLVED, INFORMATION MUST BE PROVIDED TO THE BSR OFFICE BY STYLE, SIZE AND COLOR.

17-930-061 (REV. 6/91)

FIGURE 17 Daily Exception Selling Price Report

```
                         DAILY EXCEPTION SELLING PRICE REPORT           DATE  OCT 09,YY        PAGE: 1
                                                                        TIME   9.10 AM

DATE OF SALE: OCT 08, YY
MINIMUM PRICE DIFFERENCE REQUIRED: 10.00%

STORE  SALES BILL  CLERK  STYLE  SEASON     DESCRIPTION        COLOR  SIZE  QTY  -- PRICE PER UNIT --   TOTAL EXTENDED    %
NO.      NUMBER                                                                  CURRENT     ACTUAL      DIFFERENCE     M/DOWN
-----------------------------------------------------------------------------------------------------------------------------
 001     103256     0018   SW50A   M     Cotton V-Neck Sweater  030   -   S   1   45.00       32.00        13.00         29
         103273     0026   SK1715  M     Cotton Pants           030   -   L   2   64.00       50.00        28.00         22
                                                                                ------------------       ----------    -----
                                                                         STORE DAILY TOTAL                 41.00         24

 002     204312     0042   SC1210  M     Wool Sweater           010   -   M   1   95.00       75.00        20.00         21
         204317     0042   200XT   M     2 Pces Dress Suit      020   R - 40   1  440.00      390.00       50.00         11
                                                                                ------------------       ----------    -----
                                                                         STORE DAILY TOTAL                 70.00         13


                                                                       APPROVED BY: _____
```

© 1993 Richter Management Services, Inc.

FIGURE 18 Sales and Productivity Report by Store by Salesperson

```
                         SALES AND PRODUCTIVITY REPORT BY STORE BY SALESPERSON        DATE OCT 09,YY   PAGE 1
                                        WEEK ENDING OCT 08,YY                         TIME 9:44 AM

REGION: 001 East Coast
STORE : 001 Boston

SALES PERSON   DOLLAR   DOLLAR    DOLLAR     %    UNITS    %    UNITS  DOLLARS AVG.SALE  GROSS    EMPL   HRS   %WAGES  SALES EMP $
NO    NAME     SALES    RETURNS  NET SALES STORE  SOLD   STORE /SALE   /SALE   /UNIT   PROFIT   WAGES WORKED /SALES  /HOUR PURCH
-----------------------------------------------------------------------------------------------------------------------------------
0018 George Brown
THIS WEEK       4,200      76      4,124    35.0    84   40.0   2.3     113      49     2,288     600    40   14.6    103     0
THIS PERIOD    17,255     430     16,825    34.5   246   27.8   1.6     109      68     8,457   2,432   162   14.5    104    300
THIS YEAR     149,765   3,450    146,315    35.3 2,826   38.8   1.7      88      52    74,620  22,565  1500   15.4     98   1450

0026 Mark Lewis
THIS WEEK       3,155      65      3,090    26.2    59   28.1   1.7      88      52     1,623     495    40   16.0     77    105
THIS PERIOD    12,112     255     11,857    24.3   211   23.8   1.9     106      56     6,260   2,150   160   18.1     74    105
THIS YEAR     107,270   2,380    104,890    25.3 1,687   23.2   1.3      81      62    51,396  18,915  1525   18.0     69    600


TOTAL STORE
THIS WEEK      12,465     216     12,249   100.0   210  100.0   2.1     122      58     5,980   1,702   180   13.9     68    352
THIS PERIOD    50,981   1,250     49,731   100.0   886  100.0   1.8     101      56    25,874   8,225   715   16.5     70    625
THIS YEAR     425,236  10,205    415,031   100.0 7,281  100.0   1.6      91      57   182,870  56,111  6864   13.5     60   3584


TOTAL REGION
THIS WEEK      45,243     752     44,491            983          1.6      72      45    21,885   6,583   692   14.8     64   1125
THIS PERIOD   202,853   5,223    197,630          4,184          1.7      80      47   103,882  32,581  2835   16.5     70   2152
THIS YEAR   1,602,356  42,347  1,560,009         31,483          1.4      70      50   678,401 223,004 27154   14.3     58  12003


TOTAL COMPANY
THIS WEEK     124,725   2,850    121,875          2,722          1.6      72      45    58,745  19,850  2102   16.3     58   3002
THIS PERIOD   608,124  14,987    593,137         12,584          1.6      75      47   312,715  97,100  8525   16.4     70   6100
THIS YEAR   4,707,453 108,401  4,599,052         91,123          1.4      71      51  2001,453 658,343 80131   14.3     57  35111
```

© 1993 Richter Management Services, Inc.

5. Daily Sales Reports

Today's automated business systems offer sophisticated reporting for processing and auditing sales. In addition to full accounting sales summaries, Figures 17 and 18 show information that can be provided to improve sales performance. Figure 17, DAILY EXCEPTION SELLING PRICE REPORT, highlights daily sales for items sold at prices other than the current retail (e.g., markdown price). Salesperson performance is analyzed by the use of a SALES AND PRODUCTIVITY REPORT illustrated by Figure 18.

6. Employee Discounts

As described in Unit II, an employee discount is the common practice in retail stores that allows store employees a percentage off the retail price when making purchases for themselves. It is essential to record the difference between the retail price and price paid by the employee. The procedure and form of record used for this transaction vary widely from store to store. Typically, employee discounts are listed under "Retail Reductions", but are classified separately from markdowns even though they are reductions in retail price.

D. Finding the Cost of Merchandise Sold and the Gross Margin with the Retail Method of Inventory Valuation

The retail method of inventory was introduced in department stores because it allowed a more simplified method to constantly monitor the all-important gross margin figure. The retail method of inventory eliminated a whole system of records formerly necessary to determine the valuation of an inventory at cost. Because the calculation of profit depends on cost data, the subsequent steps shown are taken in the calculation of a book inventory, and in the determination of a continual gross margin figure. The following problem illustrates the calculation of gross margin on stock plus purchases (i.e., total merchandise handled):

PROBLEM:

A junior sportswear buyer wants to calculate the gross margin figure to ascertain whether the department is "on target" and will achieve the planned gross margin goals for the season. The available season-to-date information is:

	Cost	Retail	MU%
Opening Inventory		$100,000	51%
New Purchases and Freight	$240,000	500,000	

Step 1: Begin with a retail opening inventory figure: $100,000. The opening inventory at retail ($100,000) was determined when the stock-on-hand was physically counted at the end of the previous accounting period. The cumulative markup of 51% was achieved.

Step 2: Determine a cost opening inventory figure:

$100,000 × (100% − 51%)

= $100,000 × 49%

= $ 49,000

Step 3: All new purchases ($240,000 cost, $500,000 retail) are added to the opening inventory figures ($49,000 cost, $100,000 retail) to find total merchandise handled (TMH) figures ($289,000 at cost and $600,000 at retail) with a 51.8% MU.

Step 4: The sum $475,000 ($425,000 (total of net sales) + $45,000 (markdowns)+ $5,000 (shortages) is subtracted from the retail figure of TMH ($600,000) to find the retail closing inventory figure ($125,000).

Step 5: Determine a cost closing inventory figure:

$125,000 \times (100\% - 51.8\%)$

$= \$125,000 \times 48.2\%$

$= \$ 60,250$

Step 6: The closing inventory at cost ($60,250) is subtracted from TMH at cost ($289,000) to find cost of merchandise sold ($228,750).

Step 7: The cost of merchandise sold ($228,750) is then subtracted from the net sales ($425,000)[4] to find the merchandise margin ($196,250).

Step 8: The cash discounts ($13,000) are added to the merchandise margin ($196,250), which equals $209,250.

Step 9: The workroom costs ($1,000) are subtracted from $209,250 to find the gross margin ($208,250).

FIGURE19 Calculating Gross Margin on Stock Plus Purchases

	Cost	Retail	% of Sales	Cum. MU %
Opening Inventory	$ 49,000 ↓ $\left[\begin{array}{l}\$100,000 \times (100\% - MU\%) \\ \$100,000 \times 49\%\end{array}\right]$	$100,000		51.0%
(Plus) New Purch. and Freight Total Mdse. Handled	+240,000 / 289,000	+500,000 / 600,000		51.83%
Minus the sum total of: — Net Sales — Markdowns — Shortages	+425,000 + 45,000 + 5,000 (Total Deductions) $475,000	−475,000	100.00% 10.58% 1.17%	
(Minus) Closing Inventory	−60,250 ↓ $\left[\begin{array}{l}\$125,000 \times (100\% - MU\%) \\ \$125,000 \times 48.2\%\end{array}\right]$	125,000		
Gross Cost of Mdse. Sold	228,750 ↓ ($289,000 − 60,250)			
Margin ↓ (Net Sales $425,000 − Gross Cost of Mdse. Sold 228,750)		196,250	46.2%	
(Plus) Cash Discounts		+ 13,000		
		$209,250		
(Minus) Workroom Costs		− 1,000		
GROSS MARGIN		$208,250	49%	

---

[4]In the calculation of a maintained markup, the margin on sales is determined before making adjustments for cash discounts earned and alteration costs.

From an accounting viewpoint, in the calculation of gross margin, the cash discounts and workroom costs are adjusted after the margin on the merchandise itself (i.e., maintained markup) is determined. Nonetheless, because merchants frequently negotiate cash discounts or influence the workroom factor, their impact on the gross margin must be considered.

E. The Relationship of Profit to Inventory Valuation in the Retail Method of Inventory

The value placed on an inventory has a decided effect on profits. Earlier in this unit there was a detailed examination of the mathematical calculations and records adopted by departmentalized retailers who use the Retail Method of Inventory to establish a continuing gross margin figure and to verify if a profit has been achieved. By illustrating the relationship between sales volume, cost of merchandise sold, given expenses, and the operating profit, the example that follows shows the application of the data collected through this method of inventory valuation. (For ease of comprehension of this system, the same figures are used as in the preceding calculation of gross margin to the operating profit.)

Net Sales	Cost	Retail	MU%
		$425,000	100%
Opening inventory	$ 49,000		
+ Purchases	+240,000		
Total mdse. handled	$289,000		
− Closing inventory	− 60,250		
Gr. cost of mdse. sold	$228,750		
− Cash discounts	− 13,000		
	$215,750		
+ Workroom	+ 1,000		
Net cost of mdse. sold	$216,750	−216,750	51%
Gross margin		$208,250	49%
− Operating expenses		−191,250	−45%
Net profit		$ 17,000	4%

1. On July 15, a statistical inventory indicates an on hand retail stock of $64,250. A physical count on that date reveals a stock of $62,875. What is the opening retail inventory figure for the period commencing July 16?

2. Opening inventory at retail for an outerwear department is $175,000. Purchases for the following six-month period are $490,000, net sales are $400,000, markdowns are $30,000, returns to vendors are $ 10,000, transfers (transfers out) to the third floor boutique are $15,000, and employee discounts are $6,000. Find the retail book inventory at the end of this six-month period.

3. Using the following figures from an accessories department, find the closing book stock at retail.

Physical inventory January 15	$ 85,000
Purchases January 16 through July 15	165,000
Gross sales	170,000
Returns from customers	30,000
Returns to vendors	5,000
Markdowns	10,000

4. The following figures are from a junior sportswear department:

Markdowns	$ 12,000
Purchases (retail)	315,000
Returns to vendors	20,000
Transfers in	8,000
Transfers out	4,000
Net sales	265,000
Opening inventory (retail)	180,000

(a.) Determine the closing book inventory for the period at retail.

(b.) How would the opening retail inventory for the coming six-month period be determined?

5. The following figures are from a small boutique, which has a 49% MU:

Opening inventory at retail	$16,000
Net sales	31,000
Markdowns	2,000
Purchases (retail value)	40,000

(a.) Determine the retail book inventory for the period.

(b.) Convert the closing retail inventory figure to the cost value.

*6. Distinguish between physical inventory and statistical inventory. Which one is more likely to be affected by human error? Why? Which one has become more accurate since the advent of EDP capability?

_____

*For research and discussion.

7. A lingerie department buyer was given the following data:

	Cost	Retail
Opening inventory	$ 320,000	$ 530,000
Purchases	1,118,000	2,271,000
Net sales		1,400,000
Markdowns (incl. empl. disc.)		90,000

Calculate:

(a.) The closing book inventory at retail.
(b.) The closing book inventory at cost.

8. The athletic footwear department had a closing book inventory, at retail, of $400,000 and had achieved a 53.5% markup on total merchandise handled. Determine the closing inventory at cost.

9. Find the closing inventory, at cost, of a furniture department, if:

	Cost	Retail
Net sales		$330,000
Opening inventory	$150,000	325,000
Markdowns		25,000
Returns to vendors	13,000	18,000
Employee discounts		6,500
Gross purchases	200,000	390,000

10. Using the following records from Branch Store #5, find the gross margin in dollars and percentages:

Opening inventory:	$ 156,000 at cost
	300,000 at retail
New purchases:	780,000 at cost
	1,500,000 at retail
Net sales	1,275,000
Markdowns	135,000
Shortages	15,000
Cash discounts	39,000
Alterations & workroom costs	3,000
Cumulative MU %	48%

11. Utilize the following figures to calculate:
   (a.) The closing book inventory.
   (b.) The cost value of closing inventory.

	Cost	Retail
Opening inventory	$390,500	$ 774,000
Gross purchases	690,000	1,360,000
RTV	6,400	12,000
Freight	3,260	
Gross sales		1,117,000
Customer returns		25,000
Markdowns		93,000

III. SHORTAGES AND OVERAGES

Physical inventories at current retail prices are taken at the end of the accounting period. At this same time, the "book stock" at retail is adjusted to agree with the dollar value of the physical count. Any discrepancy between the dollar value of the "book stock" and the dollar value of stock determined by the physical count of merchandise on hand is classified as a SHORTAGE (or shrinkage) or an OVERAGE. As previously described, shortages exist if physical inventory is lower than book inventory; overages exist if the physical count exceeds the statistical tally.

It is almost impossible to run a merchandising operation with 100% accuracy. Shortages or overages nearly always result, and actually are expected to occur. The shortage or overage is commonly expressed as a percent of the net sales. Regardless of the cause, the inventory shortage is fundamentally the buyer's or department manager's problem and responsibility. To keep discrepancies to a minimum is one of the many challenges a merchant must face. For internal control purposes, it is sometimes desirable to estimate shortages. This estimate is also expressed as a percent of net sales. Furthermore, even though merchandise planning is devised with estimated planned shortages in mind, generally the actual shortage exceeds the expected shortage.

Figure 23, a shortage report, which is usually calculated at the end of the accounting period, shows typical shortage information. From the data gathered in this report, a multi-store operation can pinpoint prevention, causes, and shortage remedies and can attempt to improve the shortage results.

A. Causes of Shortages and Overages

Shortages may stem from inaccurate record-keeping and/or faulty physical counts. A principal cause of shortages is pilferage which, realistically, can never be prevented completely. Overages, however, can only be caused by faulty record-keeping.

FIGURE 20 Shortage Report

DIV 10 STORE	SHORTAGES IN DOLLARS CURRENT SEASON	CURRENT -1	CURRENT -2	SHORTAGES IN PERCENTS CURRENT SEASON	CURRENT -1	DATE CURRENT -2
00						
01	8,731	2,395	3,920	2.8	0.8	1.3
06	891-	717	1,712	0.5-	0.5	1.1
09	999-	867-	1,583	0.9-	0.9-	1.5
12	293-	668	1,518	0.3-	0.8	1.6
14	5,107	507	524	5.0	0.6	0.6
15	1,056	1,361	1,928	0.8	1.2	1.6
DIV 10	12,711	4,781	11,185	1.4	0.6	1.3

The common causes of shortages and overages are:

▶ Clerical errors in the calculation of the book and/or physical inventory, which include:

Failure to record markdowns properly.

Incorrect "retailing" of invoices.

Errors in charging invoices to departments.

Errors in recording transfers.

Errors in recording returns to vendor.

Errors in recording physical inventory.

(Please note that the computer processing of these forms has minimized these clerical errors.)

▶ Physical merchandise losses, which include:

Theft by customers and/or employees.

Unrecorded breakage and spoilage.

Sales clerks' errors in recording sales.

Overweighting.

Borrowed merchandise.

Lost or incorrect price tickets.

Sampling.

B. Calculating Shortages and Overages

1. The Physical Inventory Count as a Determining Factor in the Calculation of Shortages or Overages

CONCEPT:

Shortage (or Overage) = Closing book inventory at retail – Physical inventory

PROBLEM:

Find the shortage or overage in dollars from the following figures:

Opening inventory at retail	$22,000
Purchases at retail	17,500
Net sales	18,000
Markdowns	300
Employee discounts	600
Physical inventory, end of period	19,200

SOLUTION:

Op. inv. retail (ins)	= $22,000		
+ Purchases retail	= +17,500		
Tot. merch. handled (total ins)	= $39,500	→	$39,500
Net sales	= $18,000		
+ Markdowns	= + 300		
+ Employee discounts	= + 600		
Total deductions (outs)	= $18,900	→	–18,900
Book inventory retail	=		$20,600
– Physical inventory	=		–19,200
Dollar Shortage	=		$ 1,400

2. Expressing the Amount of Shortages or Overages for a Period as a Percentage of the Net Sales for the Same Period

CONCEPT:

$$\text{Shortage \%} = \frac{\text{\$ Shortage}}{\text{\$ Net sales}}$$

PROBLEM:

For the period under consideration, the net sales of Dept. 23 are $100,000. The physical count revealed a $5,000 shortage. What was the shortage percentage for this period?

SOLUTION:

$$\frac{\text{\$ Shortage}}{\text{\$ Net sales}} = \frac{\$ 5,000}{\$100,000}$$

Shortage = 5%

3. Estimating Shortages that are Expressed as a Percentage of the Planned Net Sales Figure for Internal Control Purposes

CONCEPT:

Estimated dollar shortage = Estimated shortage percentage × Planned net sales

PROBLEM:

The seasonal plan for a department showed planned sales of $350,000 with a planned shortage of 2.5%. What was the planned dollar shortage?

SOLUTION:

$350,000 (Net sales)

×   .025 (2.5% Planned shortage)

Planned Dollar Shortage = $8,750

 ## IV. AN EVALUATION OF THE RETAIL METHOD OF INVENTORY

A. The Advantages of the Retail Method of Inventory

The benefits of the retail method of inventory are that:

- ▶ It permits control over profit because the figures for markup obtained (i.e., the difference between the cost and the retail of the total merchandise handled) and markdowns taken (upon which the realized gross margin depends) frequently are available and immediate action can be taken to protect the desired profit margin.

- ▶ It simplifies the physical inventory process because the physical inventory is taken at retail prices, which is less difficult and less expensive. Additionally, because all entries are made rapidly and no decoding is necessary, the personnel used does not require special training or experience.

- ▶ It provides a book inventory and, therefore, discrepancies (i.e., shortages and/or overages) in stock can be determined, shortage causes may be discovered, and preventive/corrective measures can be taken.

- ▶ It provides an equitable basis for insurance and adjustment claims.

B. The Limitations of the Retail Method of Inventory

The disadvantages of the retail method of inventory are that:

► It is a system of averages and therefore does not provide a precise cost evaluation of the inventory at its present cost price. This figure (i.e., cost evaluation of inventory) is calculated by applying the markup complement percentage to the retail value of the inventory. This may result in a figure that is either greater or smaller than the invoice cost of the merchandise currently received. This is the most significant weakness of this method.

► It depends upon extensive record-keeping for system accuracy.

► It is essential that all price changes be recorded.

12. A costume jewelry department showed the following figures for a six-month period:

Net sales	$125,000
Purchases (at retail)	105,000
Opening retail inventory (Feb. 1)	64,000
Markdowns	9,000
Employee discounts	2,600
Physical count (July 31)	31,000

(a.) What was the shortage in dollars?

(b.) What was the shortage in percent?

(c.) If the planned shortage was estimated at 2%, was the actual shortage more or less? By how much in dollars? In percent?

13. Find the shortage or overage percent if:

Net sales	$137,000
Opening inventory (retail)	140,000
Markdowns	7,000
Employee discounts	1,000
Retail purchases	96,000
Closing physical inventory	89,150

14. Last year, the net sales in a home fashions department were $365,000. The book inventory at year-end was $67,500, and the physical inventory was $66,000. What was the shortage percent?

15. Find the shortage or overage percentage using the following data:

Opening inventory (retail)	$204,000
Net sales	342,000
Vendor returns	4,000
Transfers to branches	8,000
Employee discounts	1,000
Purchases (at retail)	495,000
Markdowns	46,000
Closing physical inventory	287,000

16. If the retail book inventory at the close of the year is $1,500,000 and the physical inventory totals only $1,275,000, what will be the shortage percent, if net sales were $15,000,000?

17. The merchandise plan for Fall shows planned sales of $35,000 with an estimated shortage of .7%. What are the planned dollar shortages for Fall?

18. A new shop owner was reviewing figures with the store's accountant. Net sales for the first three months of business were $87,000 and the book inventory was $72,000. It was noted that the physical inventory was 2.5% lower than the book inventory. Find the shortage percentage for this three-month period.

19. For the six-month period ending in January, your department showed the following figures:

Opening inventory (retail)	$262,000
Customer returns	10,000
Returns to vendor	6,200
Employee discounts	3,800
Gross sales	910,000
Retail purchases	870,000
Markdowns	30,000
Transfers in	5,100
Transfers out	4,000
Physical inventory	170,000

(a.) What is the percentage of employee discounts?

(b.) Determine the overage or shortage in both dollars and percent.

(c.) Using a 49% markup, convert the retail opening and closing inventories to cost values.

20. The shortage in the legwear department is $3,500. This is 5% of that department's net sales. What is the sales volume of the department?

21. The coat department's net sales were $225,000, markdowns taken amounted to $15,000, and employee discounts were $3,000. The retail opening inventory for this period was $75,000, the purchases made at retail were $210,000, and the buyer estimated the shortages at 2%. Determine the estimated physical count.

22. A store with net sales of $3,500,000 has estimated its shortages to be 2%. The actual dollar shortage amounted to $72,000.

    (a.) Was this shortage higher or lower percentage than anticipated? By how much?
    (b.) What was the dollar difference between the estimated and the actual shortage?

23. The net sales of a lingerie department were $295,000; inventory on February 1 was $150,000; markdowns were 8% of net sales; purchases for this period were $362,000; the physical inventory taken July 31 was $188,400.

   (a.) Was there a shortage for this period?

   (b.) What was the shortage or overage percent for this period?

*24. Describe in detail the various methods that a merchant might use to reduce excessive departmental shortage.

*25. One of the major duties of any merchant is to control inventory discrepancies (e.g., excessive shortages or overages). Prepare a brief fact sheet for new assistant buyers that outlines the actions a merchant at the departmental level can take to accomplish effectively this responsibility. Briefly explain each action mentioned.

---

*For research and discussion.

26. A men's outerwear department had an opening inventory of $340,000. The net purchases were $78,000, gross sales were $140,250, customer returns were $11,150, and markdowns — including employee discounts — were $4,800. Shortages of 1.2% were estimated. Calculate:

(a.) The closing retail book inventory.

(b.) The estimated physical inventory.

27. For the Spring Season, the hosiery department had net sales of $700,000. On July 31, the physical inventory was $174,220 and the retail book inventory was $185,220.

(a.) What was the shortage percent for the season?

(b.) If a 2% shortage was estimated, was the actual shortage percent higher or lower than anticipated? By how much?

28. The data supplied to the activewear department in Branch Store 03 included:

	Cost	Retail
Inventory Feb. 1	$288,000	$600,000
Purchases	141,787.50	298,500
Transfers in	1,920	4,000
Freight	500	
Net sales		520,000
Retail reductions		2,400
Physical inventory July 31		360,000
Operating expenses	249,120	

Find:

(a.) The ending book inventory at retail.

(b.) The stock shortage percent.

(c.) The cost of merchandise sold.

(d.) The gross margin in dollars.

(e.) The net profit percent.

Retail Method of Inventory

Two years ago, Ms. Carol, the misses sportswear buyer, agreed with management that a separate petite sportswear department should be created. Because of increasing sales, she felt that petite sportswear had outgrown its status as a classification and deserved, in its own right, to become a separate department. Ms. Carol continued as the buyer for the newly created department, and with her enthusiastic and skillful attention, the impressive sales increases continued for the first year. The second year, however, the sales increases were minute. Ms. Carol now questioned if the category should remain a department or again be incorporated into the misses sportswear department. To make an appropriate judgement, she requested the following data for analysis:

The department had an opening inventory of $750,000 at retail that carried a 53.0% markup. During this period, the gross purchases of $570,000 at retail were priced with a 56.1% markup. The freight charges were $9,600. The merchandise returned to vendors amounted to $14,000 at cost, and $30,000 at retail. Transfers from the misses sportswear department were $3,500 at cost, and $7,600 at retail. Transfers to the misses sportswear department were $8,000 at retail, with an agreed cost of $3,700. The gross sales were $720,000; customer returns and allowances were $30,000. The markdowns taken were 12%, and employee discounts were 1%.

As Ms. Carol determined the gross margin achieved by the petite sportswear department, she weighed this against the 46.1% gross margin of the misses department. Should this "new" department continue as a separate entity? Why? Justify your decision with a mathematical comparison between the performance of the petite vs. misses sportswear departments.

UNIT V
Dollar Planning and Control

▶ Understanding and recognition of the elements of a six-month merchandise plan.

▶ Knowledge and ability to plan sales.

▶ Calculation of changes in sales as percentages.

▶ Facility to plan stock levels, using:
    Stock-sales ratio method
    Week's supply method
    Basic stock method

▶ Calculation of GMROI.

▶ Proficiency to plan markdowns.

▶ Skill to plan purchases at retail and at cost.

▶ Calculation of open-to-buy figures.

average stock	open-to-buy
basic stock method	planned purchases
BOM stock	six-month merchandise plan
dollar merchandise plan	stock-sales ratio
EOM stock	stock turnover
GMROI	weeks supply method

Profit in retailing is determined largely by maintaining a proper proportion between sales, inventories, and prices. The merchandiser is responsible for providing an inventory that reflects customer demand and that also remains within the financial limits set by management. For each department, a sales goal in dollar amounts is forecast and the size of the inventory necessary to meet these goals is planned. A budget that coordinates these sales and stocks is called a DOLLAR MERCHANDISE PLAN. It schedules planned sales month by month, the amount of stock planned for each of these months, and the amount of projected reductions. The budget is prepared in advance of the selling period to which it applies. It typically covers a six-month period, (e.g., August 1 to January 31, and February 1 to July 31). The vital information contained in this budget permits the retail merchandiser to determine the amount of purchases required. Another essential figure in the dollar plan is the projected cumulative markup percentage (i.e., the cost minus the original retail price of the total merchandise handled). Although the dollar merchandise plans used by different stores vary considerably in scope and detail, when properly planned and administered, the sales, stocks, markdowns, purchases, and markup are the indispensable figures that should result in a satisfactory net profit for any store.

The chief purpose of planning purchases is to assist the buyer in making purchases at the proper time and in the correct amounts so that the stock level is in ratio to sales. Consequently, the dollar merchandise plan also provides a control figure called OPEN-TO-BUY. This figure represents the dollar amount of merchandise the buyer may receive during the balance of a given period, without exceeding the planned stock figure at the end of the period under consideration.

Because of the benefits that result from this process, most large and many small retailers are committed to comprehensive planning activities. The dollar merchandise system of planning and control discussed in this unit is designed to protect a store's major investment, (i.e., its inventory). Because the art of this merchandising technique is broad in scope and requires an in-depth examination, it is mandatory to include this subject matter in any study related to buying for retail. Additionally, because of the nature of the text of this book, the topics and explanations in this chapter specifically focus on the mathematics used in this process and the concepts and principles are briefly defined and discussed for better comprehension.

▶ I. SIX-MONTH SEASONAL DOLLAR PLAN

As a device for unifying merchandising operations, the objectives of the dollar plan are:

- ▶ To procure a net profit by providing an instrument that plans, forecasts, and controls the purchase and sale of merchandise.
- ▶ To research previous results to repeat and improve prior successes and to avoid future failures.
- ▶ To integrate the various merchandising activities involved in determining the purchases necessary to achieve the estimated planned sales.

Figures 21-24 are examples of typical six-month seasonal dollar plans. Although the formats differ, there are certain common features. These forms are included for familiarization with these differences and similarities. In large stores, the statistical departments furnish this kind of historical data to assist the buyers and the divisional merchandise managers in making decisions on the planned sales, stock, and markdown figures from which the required planned purchases — currently referred to as planned receipts — are calculated. Additionally, it is common for a general merchandise manager and/or a controller to contribute to the planning function.

A. The Procedure, by Element, of Dollar Planning

As a guide to merchandising, the real value of dollar planning is that the figures projected for each element reflect goals that are reasonably attainable. Because the buyer is responsible for interpreting and achieving the projected figures, it is essential that the buyer be involved in the preparation of the figures. Also, a buyer who has helped to set these guidelines will be more inclined to use them. Once completed, a sound dollar plan must be adjusted to actual conditions and results during the season under consideration.

FIGURE 21 Six-Month Merchandise Plan

SALES

STORE	FEBRUARY PLAN	FEBRUARY L Y	MARCH PLAN	MARCH L Y	APRIL PLAN	APRIL L Y	MAY PLAN	MAY L Y	JUNE PLAN	JUNE L Y	JULY PLAN	JULY L Y	TOTAL SEASON PLAN	TOTAL SEASON L Y	STORE
01	27.8	26.5	43.9	40.4	57.6	55.0	79.0	75.5	46.1	43.5	26.8	25.4	281.2	266.3	01
02	43.8	42.3	59.5	55.8	84.1	81.3	120.1	116.0	61.2	58.4	39.2	37.9	407.9	391.7	02
03	44.3	42.8	56.7	53.6	84.7	81.8	123.2	118.9	69.3	65.7	43.8	42.4	422.0	405.3	03
05	14.9	14.1	23.8	22.2	33.2	31.7	48.8	46.6	27.5	26.3	15.6	15.0	163.8	155.9	05
07	10.8	10.5	15.0	14.3	18.6	18.1	29.3	28.3	17.7	17.1	9.8	9.4	101.2	97.7	07
08	18.8	.18.1	32.1	30.3	39.1	37.7	59.4	57.2	33.9	32.6	20.3	19.4	203.6	195.3	08
09	10.8	10.3	13.9	12.9	19.0	18.2	31.9	30.5	20.3	19.3	11.0	10.5	106.9	101.7	09
12	23.3	22.2	27.3	25.5	47.4	43.8	77.5	72.0	44.5	41.3	31.1	28.9	251.1	233.7	12
13	28.0	26.2	34.1	31.6	60.1	56.1	96.3	89.7	52.8	49.4	31.2	29.1	302.5	282.1	13
14	7.3	6.6	8.4	7.5	10.5	9.5	21.8	19.5	13.3	12.0	6.5	5.9	67.8	61.0	14
15	10.1	8.7	12.5	10.4	20.2	16.9	35.6	30.7	24.9	22.2	16.1	14.6	119.4	103.5	15
16	17.2	15.8	21.5	19.5	32.1	29.5	52.8	48.2	30.2	27.9	17.3	15.9	171.1	156.8	16
17	1.4	1.4	2.2	2.2	3.7	3.6	5.8	5.7	4.3	4.2	4.1	4.0	21.5	21.1	17
18	12.0	11.2	15.1	13.7	24.9	23.3	43.9	40.1	18.4	17.0	15.2	14.0	129.5	119.3	18
19	14.3	13.6	16.9	15.6	23.0	21.7	40.3	38.0	25.5	24.0	14.7	13.8	134.7	126.7	19
20	10.8	9.7	14.6	12.9	14.4	13.4	34.5	31.4	23.7	21.6	10.9	10.0	108.9	99.0	20
21	6.1	5.5	11.6	10.2	8.9	8.3	17.9	16.3	17.0	15.5	9.9	9.0	71.4	64.8	21
TOT	301.7	285.5	409.1	378.6	581.5	549.9	918.1	864.6	530.6	498.0	323.5	305.2	3064.5	2881.8	TOT

STOCK

STORE	FEBRUARY PLAN	FEBRUARY L Y	MARCH PLAN	MARCH L Y	APRIL PLAN	APRIL L Y	MAY PLAN	MAY L Y	JUNE PLAN	JUNE L Y	JULY PLAN	JULY L Y	TOTAL SEASON PLAN	TOTAL SEASON L Y	STORE
01	58.0	42.8	72.0	50.1	88.0	73.4	116.0	99.2	80.0	77.7	72.0	59.8	69.0	51.6	01
02	94.0	95.5	99.0	70.6	121.0	94.8	159.0	152.3	110.0	106.6	99.0	87.3	95.0	63.1	02
03	79.0	73.3	108.0	63.5	132.0	110.0	174.0	169.6	120.0	146.9	108.0	105.9	102.2	73.1	03
05	36.0	24.3	54.0	37.6	66.0	49.6	87.0	41.8	60.0	46.0	54.0	30.3	51.0	25.0	05
07	29.0	19.5	36.0	29.4	44.0	42.0	58.0	39.4	40.0	38.5	36.0	31.4	27.0	20.9	07
08	44.0	31.3	63.0	33.8	77.0	49.7	102.0	56.7	70.0	47.7	63.0	48.9	58.0	35.5	08
09	29.0	23.0	45.0	30.5	55.0	48.6	73.0	44.5	50.0	53.8	45.0	36.2	43.0	20.3	09
12	65.0	49.3	76.0	42.6	94.0	70.5	123.0	115.0	85.0	87.5	77.0	47.9	74.0	40.0	12
13	72.0	53.0	81.0	50.0	99.0	67.1	130.0	107.7	90.0	72.7	81.0	60.0	77.0	54.7	13
14	15.0	12.2	27.0	21.6	33.0	21.2	44.0	24.3	30.0	25.5	27.0	26.2	26.0	16.9	14
15	33.0	18.1	41.0	25.8	49.0	38.1	65.0	40.9	45.0	49.7	40.0	37.1	38.0	29.2	15
16	44.0	34.3	54.0	35.5	66.0	53.1	87.0	73.9	60.0	63.7	54.0	49.0	52.0	43.2	16
17	7.0	8.6	9.0	11.0	11.0	13.8	15.0	21.5	10.0	16.8	9.0	13.7	9.0	5.7	17
18	33.0	25.3	41.0	39.7	50.0	50.4	65.0	63.9	45.0	59.8	41.0	42.5	39.0	31.7	18
19	40.0	26.0	40.0	33.5	49.0	46.1	65.0	69.7	45.0	71.9	40.0	38.4	39.0	33.8	19
20	29.0	24.0	32.0	31.4	39.0	43.6	51.0	51.5	35.0	50.3	32.0	35.9	30.0	31.8	20
21	18.0	12.8	22.0	37.1	27.0	35.8	36.0	28.2	25.0	38.6	22.0	26.8	21.0	15.8	21
TOT	725.0	573.3	900.0	643.7	1100.0	907.8	1450.0	1200.1	1000.0	1053.7	900.0	777.3	850.0	592.3	TOT

	FEBRUARY	MARCH	APRIL		MAY		JUNE		JULY	TOTAL		
PL. RCPTS.	476.7	609.1	931.5		468.1		430.6		273.5	3189.5		PL. RCPTS.
*MD	125.0 126.3	70.0 52.0	113.0	117.2	285.0	264.6	172.0	195.6	45.0 32.3	810.0 788.0		*MD

TOTAL SEASON	PLAN	L Y			PLAN	L Y	DEPT. NUMBER:_____
MARKUP %	58.0	57.4		GM & DISC %	50.6	49.9	DEPT. NAME: MS. BUDGET COORDINATES
MD & ED %	26.9	27.8		TURNOVER	3.1	3.5	BUYER:_____

*(EXCLUDING ED)

FIGURE 22 Worksheet for Merchandise Plan

$(000)

SALES ($)	FEB.	MARCH	APRIL	MAY	JUNE	JULY	SPRING	AUG.	SEPT.	OCT.	NOV.	DEC.	JAN.	FALL	ANNUAL
96 PLAN							20500								
95 ACT.	1817	3396	3476	3066	3041	2070	16865	2303						2303	19168
94 PLAN	1643	3265	3207	3238	3034	1608	15996	2231	4095	3071	3161	4233	1681	18471	34467
93 ACT.	1427	2619	2550	2701	2742	1438	13478	1766	3157	2322	2112	3077	1457	13891	27368
92 ACT.	1224	2250	2065	2366	2447	1379	11730	1725	2536	1842	1847	2620	959	11530	23260
91 ACT.	1115	2018	1947	1977	1951	1170	10177	1451	2159	1412	1377	1988	833	9221	19399

SALES % CHG.	FEB.	MARCH	APRIL	MAY	JUNE	JULY	SPRING	AUG.	SEPT.	OCT.	NOV.	DEC.	JAN.	FALL	ANNUAL
96P/92A															
95A/91A	27.3	29.6	36.3	13.5	10.9	43.9	25.1	30.4	−100.0	−100.0	−100.0	−100.0	−100.0	−83.4	−30.0
94P/91A	15.1	24.7	25.8	19.9	10.6	11.9	18.7	26.3	29.7	32.3	49.7	37.6	15.3	33.0	25.9
93A/90A	16.6	16.4	23.5	14.2	12.1	4.3	14.9	2.4	24.5	26.0	14.3	17.4	52.0	20.5	17.7
92A/89A	9.8	11.5	6.1	19.7	25.4	17.9	15.3	18.9	17.5	30.4	34.1	31.8	15.1	25.0	19.9

EOM STOCK	FEB.	MARCH	APRIL	MAY	JUNE	JULY	AVG. SPRING	AUG.	SEPT.	OCT.	NOV.	DEC.	JAN.	AVG. FALL	AVG. ANNUAL
96 PLAN							10513								
95 ACT.	7449	9578	9714	8979	8518	7581	8337	10555							
94 PLAN	8971	9703	9586	8712	7410	6950	8241	11613	12270	11810	11256	9000	8431	10190	9390
93 ACT.	7204	8802	9009	8225	6474	6105	7339	7577	8098	8536	8138	5959	6538	7279	7401
92 ACT.	5389	5986	6182	6056	5261	5140	5477	6122	6120	6460	6692	5368	5551	5922	5742
91 ACT.	5349	6188	5614	4691	4069	4481	4916	4578	5285	5031	4998	4057	4323	4679	4822

EOM − WKS OF SUPPLY	FEB.	MARCH	APRIL	MAY	JUNE	JULY	TURNOVER SPRING	AUG.	SEPT.	OCT.	NOV.	DEC.	JAN.	T.O. FALL	T.O. ANNUAL
96 PLAN							1.95								
95 ACT.	9.8	13.0	15.7	14.9	13.1	10.5	2.02	13.3							
94 PLAN	12.1	13.6	16.1	15.2	12.4	9.8	1.94	14.5	17.3	18.1	17.7	14.9	12.1	1.81	3.67
93 ACT.	12.0	15.2	17.6	16.6	13.2	11.0	1.84	13.0	15.0	18.1	16.6	13.0	10.5	1.91	3.70
92 ACT.	10.8	11.2	13.0	14.0	12.3	10.9	2.14	12.8	12.6	15.9	16.2	13.6	11.4	1.95	4.05
91 ACT.	11.8	14.1	14.4	13.3	11.4	11.5	2.07	11.9	15.4	15.7	15.1	12.4	10.6	1.97	4.02

NET RECEIPTS	FEB.	MARCH	APRIL	MAY	JUNE	JULY	SPRING	AUG.	SEPT.	OCT.	NOV.	DEC.	JAN.	FALL	ANNUAL
96 PLAN															
95 ACT.	2728	5525	3611	2331	2580	1133	17909	5276						5276	23185
94 PLAN															
93 ACT.	3080	4217	2756	1918	992	1069	14032	3238	3677	2760	1714	898	2036	14323	28355
92 ACT.	2290	2846	2261	2240	1652	1258	12547	2707	2534	2183	2079	1296	1142	11942	24489
91 ACT.	2442	2857	1373	1054	1330	1581	10637	1549	2866	1158	1345	1047	1099	9064	19700

MARKDOWN ($)	FEB.	MARCH	APRIL	MAY	JUNE	JULY	SPRING	AUG.	SEPT.	OCT.	NOV.	DEC.	JAN.	FALL	ANNUAL
96 PLAN															
95 ACT.	395	380	456	773	797	448	3249	452						452	3700
94 PLAN															
93 ACT.	165	533	342	628	743	122	2533	314	488	479	578	996	31	2885	5418
92 ACT.	250	233	294	228	649	162	1816	253	365	489	362	1002	90	2560	4376
91 ACT.	216	216	350	252	504	292	1830	168	223	315	366	450	204	1725	3555

MARKDOWN %	FEB.	MARCH	APRIL	MAY	JUNE	JULY	SPRING	AUG.	SEPT.	OCT.	NOV.	DEC.	JAN.	FALL	ANNUAL
96 PLAN															
95 ACT.	21.7	11.2	13.1	25.2	26.2	21.6	19.3	19.6						19.6	19.3
94 PLAN															
93 ACT.	11.6	20.3	13.4	23.2	27.1	8.5	18.8	17.8	15.5	20.6	27.4	32.4	2.1	20.8	19.8
92 ACT.	20.4	10.3	14.3	9.6	26.5	11.7	15.5	14.7	14.4	26.6	19.6	38.2	9.4	22.2	19.8
91 ACT.	19.4	10.7	18.0	12.8	25.8	25.0	18.0	11.5	10.3	22.3	26.6	22.6	24.5	18.7	18.3

A13 – RTW C DRESSES
A13 – RTW C DRESSES

FIGURE 23 Merchandise Plan

		AUGUST SALES BOM STOCK		SEPTEMBER SALES BOM STOCK		OCTOBER SALES BOM STOCK		NOVEMBER SALES BOM STOCK		DECEMBER SALES BOM STOCK		JANUARY SALES BOM STOCK		SEASON SALES AVER STOCK		FEBRUARY SEASON BOM T.O. STOCK	
CHAIN	LY	365.8	576	348.8	790	306.4	704	376.5	612	310.0	607	130.9	278	1836.4	595	3.09	321
	PLAN																
T.O.	LY	.63		1.04		1.48		2.08		2.59		3.09		3.09			
	PLAN																
O.T.B.	LY																
	PLAN																
M.D.$	LY	64.8		46.3		37.8		70.8		59.2		75.1		354.0			
	PLAN																
M.D.%	LY	17.72		13.36		12.33		18.80		19.10		57.40		19.28			
	PLAN																
MU: PUR	LY	50.40		49.73		48.80		50.07		44.64		48.61		49.64			
	PLAN																
MU% S&P	LY	50.12		49.69		49.20		49.49		49.26		49.07		49.07			
	PLAN																
SHORT%	LY	2.07		2.04		2.08		2.18		2.15		2.18		2.11			
	PLAN																
G.P.$	LY	147.2		143.6		124.3		149.9		115.5		20.6		701.0			
	PLAN																
G.P.%	LY	40.24		41.40		40.55		39.80		37.25		15.76		38.17			
	PLAN																
BEG SEAS MU% STK	LY	49.80															
	PLAN																

1. Planning Sales

The planning of this figure is the most significant, and should be calculated first, because it is the basis for establishing the stock, markdown, and purchase figures. Furthermore, it is the one figure that requires the greatest skill and judgment because its accuracy depends on the detailed investigation and the careful analysis of this research.

STEP 1:

Carefully forecast future total dollar sales volume for the entire period by:

(a.) Reviewing and analyzing past sales performance for the same time period.

(b.) Considering factors that may cause a change in sales. These factors include:

▶ Current sales trends.

▶ Previous rate of growth patterns.

▶ Economic conditions.

▶ Local business conditions.

▶ Fashion factors.

▶ Influencing conditions within and from outside the store or department, (e.g., changes in store concepts, market direction, competition, etc.).

CLASS		FEB BOM	FEB SALE	FEB MD	FEB REC	MAR BOM	MAR SALE	MAR MD	MAR REC	APR BOM	APR SALE	APR MD	APR REC	MAY BOM	MAY SALE	MAY MD	MAY REC	JUN BOM	JUN SALE	JUN MD	JUN REC	JUL BOM	JUL SALE	JUL MD	JUL REC	AUG BOM	TOTAL SALE	AVE. STOCK	TURN	RETAIL	LCC
EARS R.PLAN	ACT.	533	58	42	115	548				543			120	539			126	504			127	559			212						
	P.PLAN	533	58	42	115	548	100	25	120	543	105	25	126	539	135	35	135	504	125	35	215	559	180	50	130	459	703	526	1.34	841	210
	LY	371	60	4	66	365	90	4	140	398	85	8	207	406	113	5	232	564	99	20	116	534	145	9	106	323	592	423	1.40	867	217
STRAN R.PLAN	ACT.	268	31	2	61	296				276			37	268			55	268			84	296			73						
	P.PLAN	268	31	2	61	296	50	10	40	276	55	10	57	268	70	15	85	268	60	10	98	296	90	20	64	250	356	275	1.30	405	101
	LY	301	34	0	0	256	48	2	47	245	50	2	83	259	69	0	114	282	61	2	106	308	89	0	68	241	351	270	1.30	418	105
FASHI BASIC R.PLAN	ACT.	170	21	3	93	239				208			24	193			40	182			53	202			80	275					
	P.PLAN	170	21	3	93	239	45	10	24	208	45	10	40	193	55	10	54	182	50	10	85	207	75	15	66	183	291	197	1.47	362	91
	LY	119	18	3	9	103	36	3	59	118	31	3	91	165	46	4	130	229	38	6	32	206	50	3	34	119	219	151	1.45	355	89
COLO R.PLAN	ACT.	114	17	4	0	93				190			117	188			23	148			0	124			16						
	P.PLAN	114	17	4	0	93	15	5	117	190	20	5	23	188	30	10	0	148	30	10	0	124	35	10	20	99	147	137	1.08	176	44
	LY	114	24	10	59	143	34	3	135	240	39	4	55	248	49	16	55	230	39	16	40	266	52	6	47	266	237	215	1.10	415	104
GOLD R.PLAN	ACT.	170	8	2	0	160				204			79	218			43	188			0	168			16						
	P.PLAN	170	8	2	0	160	25	10	79	204	25	5	44	218	20	10	0	188	20	10	0	168	40	15	30	143	143	179	0.80	168	42
	LY	26	5	0	42	58	16	1	41	81	20	1	107	160	11	14	22	145	11	14	22	223	43	2	43	116	114	116	0.99	361	90
FASHI R.PLAN	ACT.	172	26	3	118	261				284			102	424			214	412			60	397			64						
	P.PLAN	172	26	3	118	261	60	15	98	284	50	20	210	424	55	20	68	412	55	20	68	397	75	20	40	322	326	325	1.00	594	149
	LY	301	33	6	38	291	59	8	122	301	52	10	86	304	61	11	112	315	42	12	112	283	42	11	16	224	289	288	1.00	418	105

FIGURE 24 Total Corporate Six-Month Plan

(c.) Establishing, for the season, a percentage of estimated sales change. This is done after the past sales performance and the current conditions that cause sales changes have been reviewed and analyzed. Subsequently, the total dollar sales volume for the entire period can be calculated. The following steps describe this common procedure.

a. Calculating a Total Planned Seasonal Sales Figure When Last Year's Sales and the Planned Percent of Increase are Known.

CONCEPT:

$$\text{Seasonal planned sales} = \text{Last year (LY) sales} \times \text{Planned increase \%}$$
$$= \text{Dollar increase}$$
$$= \text{LY sales} + \text{Dollar increase}$$

PROBLEM:

If last year's seasonal sales are $1,834,900 and there was a planned 9% sales increase, what are the planned seasonal sales for this year (TY)?

SOLUTION:

Seasonal planned sales

$$= \$1,834,900 \text{ LY sales}$$
$$\times \quad 9\% \text{ Sales increase}$$
$$= \$ \ 165,141 \text{ Sales increase}$$

$$\$1,834,900 \text{ LY sales}$$
$$+ \ 165,141 \text{ Sales increase}$$

TY Seasonal Planned Sales $= \$2,000,041$[1]

b. Calculating the Percent of Sales Increase or Decrease When Last Year's Actual Sales and This Year's Planned Sales are Known.

CONCEPT:

$$\text{Percent sales increase} = \text{TY planned sales}$$
$$- \text{LY actual sales}$$
$$= \text{Sales increase}$$
$$= \frac{\text{Sales increase}}{\text{LY actual sales}}$$
$$= \% \text{ Sales increase}$$

PROBLEM:

If last year's actual sales were $1,834,900 and this year's planned sales are $2,000,000, what is the percent of sales increase?

SOLUTION:

Percent sales increase

$$= \$2,000,000 \text{ Pl. sales}$$
$$-1,834,900 \text{ Actual LY sales}$$
$$= \$ \ 165,000 \text{ Sales increase}$$

$$\frac{\$ \ 165,000 \text{ Sales increase}}{\$1,834,900 \text{ LY sales}}$$

$$= .089 \text{ or } 9\%$$

Sales Increase Percentage $= 9\%$

---

[1]On the actual plan, the seasonal planned sales would be projected at $2,000,000.

There are circumstances when the sales volume of a department or a classification is reduced because of a decreased demand. The concept for a decrease in sales is the same as for an increase of sales.

PROBLEM:

The sales of the boot classification declined significantly because of the current fashion emphasis. Consequently, this year, the shoe buyer planned the sales for this category at $500,000 this season although last year's actual sales amounted to $650,000. What was the planned percent of sales decrease for this classification?

SOLUTION:

Step 1:

% Sales decrease $=$ $\dfrac{\text{Actual LY sales} - \text{Pl. TY sales}}{\text{Actual LY sales}}$

$=$ $650,000 LY Actual sales
$-$ $500,000 TY Planned sales

$=$ $\dfrac{\$150{,}000 \text{ Sales decrease}}{\$650{,}000 \text{ LY Actual sales}}$

Percent Sales Decrease $= 23.1\%$

Step 2:

To set the individual monthly sales goals, use the seasonal distribution of the previous year's sales for the same period as a guideline. When adjusting the planned monthly sales increase or decrease, the three essential processes that influence a buyer's judgment are:

▶ Considering the department's past experience with respect to the normal percentage distribution of sales for the planning period.

▶ Comparing the monthly percentage distribution with industry performance.

▶ Adjusting monthly sales figures because of shifting dates of certain holidays, planned special promotions, and other merchandising strategies.

2. Planning Stocks

Because the merchandising policies of retail stores differ, there is no absolute formula for developing the variety of a stock assortment. However, the planning phase of stock investment is accomplished through the dollar plan. In the planning and control of dollar stocks, every merchandiser's objective is to:

▶ Maintain adequate assortments (i.e., reasonably complete from a customer's viewpoint).

▶ Regulate the dollar investment of stocks in relation to sales to obtain a satisfactory balance between these two factors.

▶ **For assignments for the previous section, see practice problems 1-8.**

FIGURE 25 Conventional Department Stores Monthly Distribution

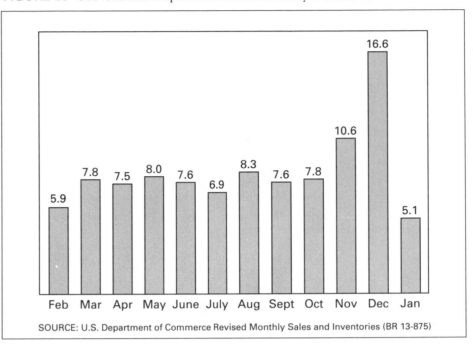

SOURCE: U.S. Department of Commerce Revised Monthly Sales and Inventories (BR 13-875)

After planned monthly sales are established, the amount of dollar stock that is required on hand at the beginning of each month (BOM stocks) and/or the end of each month (EOM stocks) must be determined. The EOM stock for a particular month is the same as the BOM stock for the following month; for example, if $230,000 is the EOM stock for February, this same figure is the BOM stock for March.

There are variations in the methods of calculating individual monthly stock figures. Before discussing these possible techniques, it is essential to examine TURNOVER because it represents the degree of balance between sales and stocks. The rate of stock turnover measures the velocity with which merchandise moves into and out of a department or store. Turnover, or rate of stock turnover, is a merchandising figure. It indicates the number of times that an average stock has been sold and replaced during a given period, the number of times goods have been turned into money, and subsequently, money turned back into goods. Despite the fact that turnover is a resultant figure, it can be planned, controlled, and for convenience of comparison, is usually expressed as an annual or semi-annual figure.

3. Determining the Turnover Figure

Every retailer should understand the importance of turnover to make better use of capital investment, to control inventories, and ultimately, to realize optimum profits. It acts as an index to efficient merchandising. Successful stock planning does not begin with turnover, but results in achieving the desired rate of stock turnover. This term indicates the number of times that an average stock is sold for a given period of time, which, unless otherwise stated, refers to a period of one year. However, turnover may be computed on a weekly, monthly, or seasonal basis. The actual number of stock turns varies with the type of merchandise and price. Generally speaking, lower price ranges turn more rapidly than higher ranges; apparel and accessories turn more rapidly than home furnishings.

Typical average turnover figures for a particular type of goods are most important as a method of comparison. The following average rate of turnover figures[2] show the range that may occur:

Misses Dresses — 3.5

Men's Clothing — 2.4

Sleepwear and Robes — 3.6

Children's Footwear — 2.2

Linens & Domestics — 1.8

Millinery — 2.7

Turnover is important to a merchandising operation because it:

▶ Stimulates sales by presenting fresh merchandise to the customer.

▶ Reduces markdowns by keeping the flow of new goods constant, thereby curbing the accumulation of large amounts of older stock.

▶ Lowers cost of goods sold because the "open-to-buy" position permits the buyer to take advantage of special prices and offerings.

▶ Decreases interest, merchandise taxes, and other operating expenses as a percentage of net sales.

The stock turnover rate can be calculated on either a unit or dollar basis, but for the purposes of this text, the dollar basis will be examined.

a. Calculating Turnover When Average Stock and Sales for the Period are Known.[3]

The dollar figures of stock turn can be determined on either a cost or retail basis. Generally, in stores that use the retail method of inventory, the rate of stock turn is determined on a retail basis. Essential for accuracy, however, is that both sales and inventory be calculated on the same foundation.

CONCEPT:

$$\text{Turnover} = \frac{\text{Net sales for period}}{\text{Average stock for same period}}$$

PROBLEM:

For the year, the infants' department had net sales of $2,000,000. The average stock during this period was $500,000. What was the rate of stock turn?

SOLUTION:

$$\text{Turnover for the period} = \frac{\$2,000,000 \text{ Net sales}}{\$\ 500,000 \text{ Average stock}}$$

Turnover   = 4

---

[2]Most current figures from MOR of Department and Specialty Stores, NATIONAL RETAIL FEDERATION. These figures represent performance at the time of publication and should not be construed to be ideal.

[3]The same method can be used to calculate either a monthly or yearly turnover figure.

b. Calculating Average Stock When Planned Sales and Turnover are Known.

CONCEPT:

Average stock = $\dfrac{\text{Planned sales for period}}{\text{Turnover rate}}$

PROBLEM:

The hosiery department planned sales of $2,000,000 with a stock turn of 4 as the goal. What should be the average stock carried for the period under consideration?

SOLUTION:

Average stock for period = $\dfrac{\$2,000,000 \text{ Pl. sales}}{4 \text{ Turnover rate}}$

Average Stock = $500,000

By understanding the relationship of the average stock, planned sales, and turnover, and by substituting the known factors of the basic formula, the unknown can be calculated, (e.g., net sales = stock turnover × average stock at retail).

c. Calculating Average Stock When Monthly Inventories are Known.

Because the determination of the average inventory directly affects the rate of stock turn, there is a need for a common method among stores and retail establishments to determine the average stock amounts so that the comparison of stock turns can be meaningful. Under the retail method of inventory, an AVERAGE STOCK is the sum of the retail inventories at the beginning of each year, season, month, or week, which is added to the ending inventory and then is divided by the number of inventories used. This is the most accurate and commonly used method because a monthly book inventory figure is available. For example, to obtain an average retail stock figure for a year, the 12 stock inventories at the beginning of each month are added to the ending inventory and the total sum is divided by 13. If the stock turnover rate is computed for a shorter period than one year, the same principle is applied, (e.g., for determining a six-month turnover rate, add the seven stock-on-hand figures and divide by 7). This turnover rate, which has been computed for a period of less than a year, can then be converted to an equivalent annual rate.

Additionally, an average stock figure can be calculated in cost as well as retail dollars. The method of finding an average monthly inventory at cost is exactly the same as determining the average stock at retail. For reasons that should be apparent, it is incorrect to mix cost figures with those at retail in the same turnover calculation, so when this type of figure is desired, only the cost inventory figures are used.

CONCEPT:

Average stock = $\dfrac{\text{Sum of beginning inventories} + \text{Ending inventory for given period}}{\text{Number of inventories}}$

PROBLEM:

Find the average stock and the turnover for this period.

	Sales		BOM Stocks
January	$10,000	Jan.1	$13,000
February	8,000	Feb. 1	12,000
March	14,000	Mar. 1	17,000
April	16,000	Apr. 1	19,000
May	12,000	May 1	15,000
June	14,000	June 1	18,000
July	10,000	July 1	14,000
August	6,000	Aug. 1	10,000
September	12,000	Sept. 1	16,000
October	11,000	Oct. 1	15,000
November	12,000	Nov. 1	15,000
December	15,000	Dec. 1	18,000
		Dec. 31	($13,000 is ending inventory)

SOLUTION:

Sum of 12 BOM stocks	=	$182,000
+ EOM stock	=	+ 13,000
	=	$195,000

Average stock = $\dfrac{\$195,000 \text{ (Sum of 13 figures)}}{13 \text{ (Number of inventories)}}$

Average Stock for One Year = $ 15,000

After calculating an average stock figure, the formula of net sales divided by the average stock is applied to determine the turnover rate. For example:

Turnover = $\dfrac{\$140,000 \text{ (Sum of 12 monthly sales figures)}}{\$ 15,000 \text{ (Average stock for the period)}}$

Turnover = 9.3

This same method is used to calculate an average stock for a shorter period.

The following problem applies the formulas for average inventory and turnover. It illustrates the determination of the average stock for a shorter period than one year, and also shows how to convert the turnover to an annual rate.

EXAMPLE:

	Sales	Stock-on-Hand (Book Inventory)
Feb. 1	$ 20,000	$ 50,000
Mar. 1	27,500	60,000
Apr. 1	35,000	75,000
May 1	32,500	70,000
May 31		60,000
Total sales	$115,000	$315,000 Sum of inventories

Average stock	$= \dfrac{\$315,000}{5}$	Sum of inventories
		Number of inventories
Average stock	$= \$ 63,000$	
Stock turn rate	$= \dfrac{\$115,000}{\$\ 63,000}$	Net sales
		Average stock
	$= 1.83$ for 4 months or 1/3 year	
Annual Rate	$= 1.83 \times 3$	
Annual Turnover Rate	$= 5.49$	

In planning stocks for a season, from the standpoint of stock turnover, merchandise is apportioned so that the average stock is related to the sales for the entire period. This approach does not offer a basis for planning a specific amount of stock to be on hand to achieve a planned sales figure.

4. Determining Gross Margin Return by Dollar of Inventory

While retailers have stock turn goals to judge the efficiency of the balance between sales and stocks, when money is tight there is a greater emphasis on the relationship between the inventory investment or working capital and its ability to produce gross margin dollars. The objective is that a minimum dollar amount invested produces a maximum amount of gross margin.

The ratio of productivity of each dollar invested in inventory is similar to a share of stock, and the productivity of this dollar is equivalent, to some degree, to earnings per share. Stock turnover influences and affects the utilization of money invested in inventory. The more times money is converted into sales and gross margin, the greater the return per dollar of inventory. Today, the frequency of this "re-investment" is an increasingly significant factor. The measurement of the efficiency of investment in inventory is referred to as GROSS MARGIN RETURN PER DOLLAR OF INVENTORY, which is commonly known as GMROI (pronounced jim-roy). This element is used by financial analysts to measure capital turnover.

Merchandisers in retailing are responsible for a successful stock turnover and, consequently, are focussed on it. Both stock turn and GMROI are involved with inventory productivity and the relationship between these two elements is that the calculation of stock turn utilizes an average inventory at retail, and GMROI uses an average inventory at cost.

The following example illustrates this concept:

Given:

Net sales	$= \$1,200,000$
Gross margin	$= \$\ \ 480,000$
Average inventory, retail	$= \$\ \ 480,000$
Average inventory, cost	$= \$\ \ 240,000$

Find:

(a.) Stock turnover	$= \dfrac{\$1,200,000}{\$\ \ 480,000}$	Net sales
		Avg. inv. at retail
	$= 2.5$	
(b.) GMROI	$= \dfrac{\$480,000}{\$240,000}$	Gross margin
		Avg. inv. at cost
	$= 2.0$	

This is the most simple level of calculating GMROI. It is apparent that as more sales and gross margin dollars are generated, without increasing inventory, GMROI will increase.

It is more practical to think of GMROI in terms commonly used in merchandising. These factors, which can be managed and influenced, are:

▶ Markup percentage.

▶ Gross margin percentage.

▶ Turnover.

Another way to think of the calculation is:

$$\frac{\text{How much is made on a sale} \times \text{How long it takes to sell it}}{\text{How much was paid for it}}$$

This results in the following calculation:

CONCEPT:

$$\frac{\text{Gross margin \%} \times \text{Turnover}}{100\% - \text{Markup \%}}$$

PROBLEM:

The year-to-date results of the intimate apparel department are:

Gross margin — 40%
Turnover — 2.5
Markup — 50%

What is the GMROI achieved by this department?

SOLUTION:

$$
\begin{aligned}
\text{GMROI} &= \frac{\text{G.M. \%} \times \text{T.O.}}{100\% - \text{MU \%}} \\[2mm]
&= \frac{40\% \text{ G.M.} \times 2.5 \text{ T.O.}}{100\% - 50\% \text{ MU}} \\[2mm]
&= \frac{.40 \times 2.5}{100\% - .50} \\[2mm]
&= \frac{1}{.50} \\[2mm]
&= 2
\end{aligned}
$$

The GMROI result is identical for either calculation but the second example shows the relationship of the components and demonstrates how future results can be improved.

▶ **For assignments for the previous section, see practice problems 9-23.**

5. Methods of Stock Planning

It is common to plan monthly stock figures for the beginning of the period by a STOCK-SALES RATIO method. This technique, illustrated in the following examples, indicates the relationship between stock-on-hand at the beginning of the month and the retail sales for the same month.

a. Setting Individual First of Month Stock Figures by Stock-Sales Ratio Method. After planned monthly sales figures are established, the amount of dollar stock that is required on hand at the beginning of each month (BOM stocks) and the end of each month (EOM stocks) must be determined. This relationship is referred to as a STOCK-SALES RATIO. Generally, the BOM stock-sales ratio is used to balance planned monthly stocks with the planned monthly sales. Standard stock-sales ratios in departments can be established by evaluating the actual past stock-sales ratio performance of the department that has proven to provide the proper relationship. Additionally, researched guidelines of data showing typical monthly stock-sales ratios are available in the MOR, published by the National Retail Federation, and this can be used as a source of information in planning monthly stock proportions. Because monthly stock-sales proportions vary, it is necessary to establish the proper ratio for an individual month. Figures 26 and 27 illustrate typical stock-sales by store volume for specialty stores and department stores.

FIGURE 26 Monthly Stock-Sales Ratios by Store Volume for Specialty Stores

Store Volume	Feb	Mar	Apr	May	June	July	Aug	Sept	Oct	Nov	Dec	Jan
Under $20 million	6.43	5.88	6.16	6.58	6.11	6.80	6.05	6.38	7.01	6.22	5.81	8.80
Over $20 million	5.65	4.63	4.81	4.79	4.06	4.92	5.15	5.04	6.18	5.42	2.72	6.01
Under $5 million	6.50	5.36	5.85	6.03	5.53	6.24	5.42	5.75	6.55	5.87	3.17	8.08
$5–$20 million	6.37	6.41	6.48	7.13	6.68	7.36	6.67	7.00	7.47	6.58	8.45	9.52
$20–$100 million	4.80	4.10	3.24	4.30	4.17	3.23	3.67	4.51	4.14	4.34	3.11	4.32
Over $100 million	5.82	4.73	5.12	4.89	4.04	5.26	5.44	5.15	6.59	5.63	2.64	6.34

FIGURE 27 Monthly Stock-Sales Ratios by Store Volume for Department Stores

Store Volume	Feb	Mar	Apr	May	June	July	Aug	Sept	Oct	Nov	Dec	Jan
Under $100 million	6.28	4.78	5.07	5.24	5.07	5.34	5.69	5.48	5.35	5.07	4.26	7.12
$100–$300 million	6.45	4.10	4.57	5.63	4.91	6.21	5.31	4.55	4.63	4.30	2.41	6.62
$300–$600 million	5.54	4.48	5.63	5.21	4.43	6.00	4.61	4.86	5.32	5.08	2.89	8.68
Over $600 million	4.87	3.83	5.03	4.85	4.14	5.11	4.62	3.73	5.18	4.32	2.92	6.43

b. Calculating Stock-Sales Ratio When Retail Stock and Sales for a Given Period Are Known.

CONCEPT:

Stock-sales ratio = $\dfrac{\text{Retail stock at given time in the period}}{\text{Sales for the period}}$

PROBLEM:

On Feb. 1, the boys' wear department had a retail stock of $120,000. The planned sales for this month were $20,000. Find the stock-sales ratio for the month of February.

SOLUTION:

Stock-sales ratio $= \dfrac{\$120,000 \text{ BOM Stock}}{\$20,000 \text{ Feb. sales}}$

Stock-Sales Ratio = 6

c. Calculating BOM Stock When Planned Sales and Stock-Sales Ratio Are Known.

CONCEPT:

BOM stock = Planned monthly sales × Stock-sales ratio

PROBLEM:

The yard goods department planned sales of $40,000 for the month of July. Past experience in the department showed an 8.2 stock-sales ratio was successful. What should be the planned BOM stock for July?

SOLUTION:

$$\begin{array}{rl} \text{BOM July stock} = & \$\ \ 40,000 \text{ Planned July sales} \\ \times & \underline{\qquad 8.2} \text{ Stock-sales ratio} \end{array}$$

BOM Stock = $328,000

▶ **For assignments for the previous section, see practice problems 24-27.**

6. Setting Stock Figures by the Weeks Supply Method

The WEEKS SUPPLY METHOD plans inventory size on a weekly basis. The set amount of stock equals a calculated number of weeks supply. The number of weeks supply that is to be "on hand" depends on the planned turnover figure to be achieved and is used as a guide to set the number of weeks inventory supply. This technique of stock planning is best used in a department that primarily carries staple merchandise and/or has a relatively stable sales volume. Please note that because the stock size is in direct relation to the planned weekly sales, it can result in an excessive stock condition at the peak selling periods or in dangerously low stocks during the slower months.

a. Calculating the Number of Weeks Supply.

CONCEPT:

Number of weeks supply = Weeks ÷ Desired turnover

PROBLEM:

Department #32 has a planned stock turnover of 4.0 for the six-month period. Determine the number of weeks supply needed to achieve the desired turnover.

SOLUTION:

Number of weeks supply $= \dfrac{26 \text{ weeks (6 month)}}{4.0 \text{ turnover}}$

Number of Weeks Supply = 6.5

b. Finding Planned Stock, Given Turnover, and Weekly Rate of Sales.

CONCEPT:

Planned stock = Average weekly sales × Number of weeks supply

PROBLEM:

A department has an average weekly sales rate of $9,800 and a planned turnover of 4.0 for the six-month period. Calculate the amount of stock to be carried.

SOLUTION:

Step 1: Find the number of weeks supply given the turnover and the supply period.

$$\text{Number of weeks supply} = \frac{26 \text{ weeks (6 month)}}{4.0 \text{ turnover}}$$

Number of Weeks Supply = 6.5

Step 2: Find planned stock given the average weekly sales and the number of weeks supply.

Planned stock = $ 9,800 Average weekly sales × 6.5 No. of weeks supply

Planned Stock = $63,700

▶ **For assignments for the previous section, see practice problems 28-31.**

7. Setting Beginning of the Month Stock Figures by Basic Stock Method

Another approach to balancing stocks and sales is the BASIC STOCK METHOD, which is the average inventory for the period. This average inventory is derived simply from the planned sales and turnover. For example, if the anticipated six-month sales total $960,000 and a turnover of 2 is desired for the season, average inventory can be calculated by applying the formula:

$$\frac{\text{Sales}}{\text{Turnover}} = \frac{\$960,000}{2} = \$480,000 \text{ Average inventory}$$

This technique presumes that the retailer will begin each month with a minimum amount of basic stock that remains constant regardless of the monthly sales to be achieved. Generally, it is best to apply this basic stock (i.e., fixed quantity of stock maintained throughout the season) when the annual turnover rate is 6 or less because using it with higher stock-turn merchandise would result in unrealistic basic stocks.

a. Calculating the Average Monthly Sales.

	Planned Sales
February	$110,000
March	150,000
April	160,000
May	180,000
June	210,000
July	+150,000
Total Sales	$960,000 ÷ 6 = $160,000 Average monthly sales

b. Finding a Basic Stock Given the Average Monthly Sales and the Average Inventory. The average inventory minus the average monthly sales produces a basic stock at retail. For example:

Average inventory	=	$480,000
– Average monthly sales	=	–160,000
Basic Stock	=	$320,000

c. Finding a BOM Stock When a Basic Stock and Planned Sales for the Month are Known. The calculated basic stock figure is added to each month's planned sales to determine the BOM amount for the month.

CONCEPT:

BOM stock = Basic stock + Planned sales for the month.

PROBLEM:

In an accessories department that has planned a turnover of 4 for the Fall season, the estimated sales are as follows:

August —	$28,000	November —	$36,000
September —	30,000	December —	40,000
October —	32,000	January —	26,000

Calculate the BOM stocks using the basic stock method.

SOLUTION:

Step 1: Determine average monthly sales.

August	—	$ 28,000
September	—	30,000
October	—	32,000
November	—	36,000
December	—	40,000
January	—	+ 26,000
Total for the season	—	$192,000 ÷ 6 = $32,000

Average Monthly Sales = $32,000

Step 2: Calculate average inventory.

Average Inventory = $192,000 Sales ÷ 4 Turnover = $48,000

Step 3: Find basic stock.

Basic stock at retail	=	Average inventory	$48,000
		– Average monthly sales	–32,000
		Basic Stock	= $16,000

Step 4: Calculate BOM stock by adding basic stock to monthly planned sales.

BOM STOCK =	*Monthly Sales*			*Basic Stock*		*BOM Stock*
	August	$28,000	+	$16,000	=	$44,000
	September	30,000	+	16,000	=	46,000
	October	32,000	+	16,000	=	48,000
	November	36,000	+	16,000	=	52,000
	December	40,000	+	16,000	=	56,000
	January	26,000	+	16,000	=	42,000

▶ **For assignments for the previous section, see practice problems 32-35.**

8. Planning Markdowns

In dollar planning it is important to include markdowns because they reduce the total value of the stock available for sale. Careful planning of markdowns helps reduce the amount of markdowns taken and this helps to increase the net profit figures. The amount of planned markdowns to be taken is expressed both in dollars and as a percentage of planned sales. Because the dollar markdowns taken vary greatly, the percentage data is more significant for comparison of past and present performance. The percentage of markdowns taken also varies with different lines of merchandise, different months, and different seasons. The planned markdown figure is usually based on a normal amount determined from past experience. Typical markdown percentages are available for industry comparison, as illustrated in Figure 28.

FIGURE 28 Typical Markdown Percentages (Including Employee Discounts)

Typical Markdown Percentages (Including Employee Discounts)[4]			
Women's Footwear—	24.5%	Boy's Clothing—	29.0%
Fine Jewelry & Watches—	7.2%	Small Appliances—	13.3%
Infants & Todlers—	23.6%	Cosmetics & Toiletries—	1.1%

Step 1: Set the total markdown amount (stated as a percentage of total season's sales) for the entire period by:

(a.) Reviewing and analyzing past markdown performance for the same period and for the entire period under consideration.

(b.) Considering factors that may effect a change in markdowns.

Step 2: Convert the planned markdown percentage of sales to a total dollar figure for the season.

Step 3: Apportion total dollar planned markdowns by month. (Please note that the distribution of monthly markdown goals does not mean necessarily that markdowns and sales will be in the same proportion during each of the months of the season.)

EXAMPLE:

Last year, the net sales for the season in the swimwear department were $100,000. The total amount of markdowns taken during the entire period totaled $5,000. The buyer, upon reviewing the past performance of the department and in preparation for planning markdowns for the same period this year, decided the $5,000 markdown figure taken previously was normal and compared favorably with standard markdown percentages established for this type of merchandise. Consequently, last year's dollar markdown amount was converted to a percentage:

$$\frac{\text{Last year's markdown}}{\text{Last year's net sales}} \quad \frac{\$\ \ 5,000}{\$100,000} = 5\% \text{ Markdowns}$$

---

[4]MOR of Department Specialty Stores in 1990, National Retail Federation.

The total planned sales figure established by the buyer for the season under consideration was set at $110,000. Because a repetition of the percentage of markdowns was desired for the forthcoming season, the dollar amount of markdowns to be taken was determined by:

$110,000 Planned sales × 5% Planned markdowns = $5,500 Planned total markdown for the season

$5,500 would then be apportioned to the individual months of the period.

9. Planning Markups

Although markup planning and calculations are treated in depth in Unit III, it is necessary to add here that after the initial markup has been carefully planned, the buyer will have to constantly manipulate the actual markup to date in relation to the markup on additional purchases for the season to obtain the planned seasonal markup percentage. To "protect" profitability, the gross profit figure, including an estimated shortage amount, can be calculated easily as in the following example, which shows the facts that should be considered:

(1.) For the Spring season, a department has planned:
- ▶ Sales   @ $2,000,000.
- ▶ Markdowns @ $ 200,000 (10%).
- ▶ Shortages   @ $ 40,000 (2%).

(2.) The department came into the period under consideration with an opening inventory at retail of $500,000, and a cumulative markup for the period of 52%.

(3.) A closing inventory of $500,000 at retail is projected for the end of the Spring period.

(4.) Management projects a desired gross margin of 48.5% for the Spring period.

(5.) Based on a 48.5% gross margin, the cumulative markup (stock + purchases) is calculated:

$$\text{Cumulative MU\%} = \frac{48.5\% \text{ G.M.} + 12\% \text{ Red. (10\% MD's} + 2\% \text{ Shtg.)}}{100\% \text{ Sales} + 12\% \text{ Reductions}}$$

$$= \frac{60.5}{112}$$

$$= 54\%$$

(6.) The markup percentage required on new purchases to meet desired goals is determined by the following method:

Total Mdse. Needs	Retail	Cost	MU%
$2,000,000 Net sales	$2,740,000	$1,260,400	54%
		↓	
+ 200,000 Markdowns		($2,740,000 × 46%)	
+ 40,000 Shortages			
+ 500,000 Cl. inv. at retail			
$2,740,000			
− Opening inventory	− $ 500,000	− $ 240,000	52%
		↓	
		(500,000 × 48%)	
New purchases	$2,240,000	$1,020,400	54.4%

MU% on new purchases is
$$\begin{array}{r} \$2,240,000 \text{ R.} \\ -1,020,400 \text{ C.} \\ \hline 1,219,600 \text{ MU} \end{array}$$

$$\frac{\$1,219,600 \text{ MU}}{2,240,000 \text{ R.}} = 54.446 \text{ or } 54.4\%$$

Given:

Dollar sales (from plan)	=	$2,000,000
Dollar markdown (from plan)	=	$ 200,000
(10%)		
Anticipated dollar shortage		
Shortage % × Total planned dollar sales)	=	$40,000 (2%)
Planned MU% on stock + Purchases	=	54.0%
Cost or complement % on planned stock + Purchases	=	46.0%

Cost of planned sales:

Pl. MD = $200,000 × .46	=	$	92,000
Pl. shortage = 40,000 × .46	=		18,400
Pl. sales = $2,000,000 × .46	=	+	920,000
Total	=		$1,030,400
Planned $ sales	=		$2,000,000
Cost of planned sales	=		−1,030,400
Planned $ gross profit	=		$ 969,600
Planned Gross Profit %	=		$ 969,600 = 48.5%
			2,000,000

10. Planned Purchases

One objective of planning is to assist the buyer in proper timing and in purchasing the correct amounts of goods. Therefore, when the planned sales, stocks, and markdowns have been determined, the amount of monthly purchases is automatically calculated by a formula. PLANNED PURCHASES refers to the dollar amount of merchandise that can be brought into stock during a given period. Store reports frequently identify these figures as receipts. (See Figures 25 and 27.) Generally, purchases are pre-planned at retail value for each month and then converted, by formula, to a cost figure by applying the planned markup.

a. Calculating Planned Monthly Purchases at Retail.

CONCEPT:

Planned monthly purchases = Pl. EOM stock

+ Pl. sales for month

+ Pl. MD's for month

= Total merchandise requirements

– Pl. BOM stock

= Pl. purchase amount

PROBLEM:

From the following planned figures for the lingerie department, calculate the planned purchase amount for June:

SOLUTION:

Planned stock June 1	$ 38,000
Planned sales for June	100,000
Planned markdowns for June	5,000
Planned stock for July 1	40,000
Planned June sales	$100,000
+ Planned EOM stock	+ 40,000
+ Planned June markdowns	+ 5,000
Total merchandise requirements	$145,000
– Planned BOM stock	– 38,000
Planned Monthly Purchases	$107,000

b. Converting Retail Planned Purchases to Cost.

CONCEPT:

Pl. purch. at cost = Pl. purch. (Retail) × (100% – Pl. MU%)

PROBLEM:

The planned retail purchases for February were $103,000, and the planned MU% was 41.5%. Calculate the planned purchase amount at cost.

SOLUTION:

Planned purchases = $103,000 Pl. ret. pur. × (100% – 41.5% Pl. MU%)

= $103,000 × 58.5%

= $60,255

Planned Purchases at Cost = $60,255

FIGURE 29 Purchase Planning

	Planned	
Planned purchases for October based on *planned* figures	$20,000	Planned Sales
	+42,000	Planned E O M Stock
	+ 500	Planned Markdowns
	$62,500	Total Merchandise Requirements
	−40,000	Planned B O M Stock
PLANNED PURCHASES =	$22,500	
	Revised	
Planned purchases for October based on *revised* figures	$22,000	Revised Sales Plan
	+42,000	Planned E O M Stock
	+ 500	Revised Markdowns
	$64,500	Total Merchandise Requirements
	−40,000	Revised B O M Stock
ADJUSTED PLANNED PURCHASES =	$24,500	

c. Adjusting the Planned Purchases. During the season, as the merchandising activities are performed, the actual results are checked against the planned figures. Sometimes this reveals the need to adjust the original planned figures for either a month or for the balance of the season, because of a deviation from the planned sales thus far or because of a change in circumstances. For example, if sales and/or markdowns are actually larger than planned, purchases must be greater than planned to achieve the level of stock planned. Conversely, if these factors are actually less than planned, a downward revision of purchases is required. Figure 32 illustrates the variation between planned figures and actual results.

NOTES

1. In the children's department, the August sales were planned at $247,500 because of additional promotional events. The actual sales for this month last year were $225,000. What is the percent increase in planned sales for August of this year?

2. For the Spring period, the handbag department's total seasonal sales volume for last year was $750,000. If there is an 8% increase this year, what is the dollar amount of sales planned?

3. During the month of August, sales for a sporting goods store are planned at $900,000, which is 15% of the planned season's total sales. Calculate the monthly sales for the balance of the season, if sales are planned as follows:

   August — 15%   November — 18%
   September — 14%   December — 25%
   October — 16%   January — 12%

4. A small retailer is planning a 10% reduction in sales due to competition from a major new department store. If last year's sales were $500,000, what sales figure should be planned for this year?

5. After a detailed analysis of sales by classification, the hosiery buyer determined a decline in the knee-hi category. Last year's sales for this category were $75,000. This year the sales for this classification are planned at $60,000. What is the percentage of sales decline?

6. Using Figures 21-24, prepare a comparative analysis by element of the six-month plan shown in each figure.

7. Seasonal sales for last year were $855,000 and this year the merchandise manager planned for a sales increase of 8%. What is the estimated planned dollar sales figure for this year?

8. If actual sales for last year were $855,000 and this year's planned sales are $923,400, calculate the percentage of sales increase for this year.

9. Determine the seasonal stock turnover on the basis of the following information:

Retail Inventory		Sales
Aug. 1	$36,000	$20,000
Sept. 1	68,000	28,000
Oct. 1	60,000	44,000
Nov. 1	40,000	28,000
Dec. 1	28,000	24,000
Jan. 1	28,000	10,000
Jan. 31	18,000	

10. A housewares department has an average inventory of $112,000 at retail, with a stock turnover of 2. What are the department sales for the period?

11. The net sales in the neckwear department were $240,000, with a stock turnover of 3. Find this department's average stock at retail.

12. What amount of average stock should be carried by a costume jewelry department with net sales of $920,000 for the year and a stock turn of 4?

13. What is the average stock of a department with annual sales of $1,350,000 and an annual stock turn of 4.5?

14. Calculate the yearly turnover using the following figures from an outerwear department:

Gross sales	$497,500
Customer returns	47,500
Inventory 8/1	85,000
Inventory 10/1	105,000
Inventory 12/1	250,000
Inventory 2/1	53,000
Inventory 4/1	90,000
Inventory 6/1	80,000
Inventory 8/1	76,000

15. For the year, a shoe department has net sales of $2,500,000. The average stock carried during this period was $1,000,000. What was the annual rate of stock turn?

16. Find the average stock from an activewear department's inventory figures:

BOM January	$ 62,000	BOM August	$ 78,000
BOM February	64,000	BOM September	78,000
BOM March	70,000	BOM October	68,000
BOM April	74,000	BOM November	64,000
BOM May	88,000	BOM December	60,000
BOM June	100,000	EOM December	62,000
BOM July	120,000		

17. Departmental gross sales for the year were $220,000 with customer returns of 10%. During the year these inventories were taken:

Date	Inventory Value (Retail)
January 1	$37,000
April 1	36,500
July 1	41,500
October 1	46,500
Jan. 1 of following year	38,500

Find:

(a.) The average stock for the year.

(b.) The annual rate of turn.

18. The sales for October were $8,000; the stock on October 1 at retail value was $24,000; the stock on October 31 was $28,000. What was the stock turn for the month?

19. The sales in the children's department for the year amounted to $416,000. The stock at the beginning of the year was $120,000 at retail, and for the end of the year, the retail stock figure was $140,000. What was the stock turn for the year?

20. The men's furnishings department had the following performance for the period under consideration:

   ▶ 26.1 Gross margin.
   ▶ 2.8 Turnover.
   ▶ 30.7% Markup.

   What GMROI was achieved?

21. The same men's furnishings buyer (in problem 20) made some spectacular purchases for an annual store event that increased the markup to 33.8%, with the other factors being constant. What GMROI resulted from this strategy?

22. Last Fall season, the active sportswear department achieved a 45.9% gross margin with a stock turn of 2.1. The markup was 60.3%. During this Fall season, the gross margin was increased to 50.5%. Compare the GMROI performance from last year with this year.

23. The budget sportswear department achieved a 26.1% gross margin and had the same 30.7% markup as the men's furnishings department, with a stock turn of 8. What was the resultant GMROI for this department?

24. What is the planned stock-sales ratio in the handbag department when beginning of the month stock is planned at $19,000 and planned sales are $9,100?

25. Planned sales in the costume jewelry department for April are $130,000 and the planned stock-sales ratio is 2.4. What should be the stock figure on April 1?

26. The innerwear department buyer decided that a stock-sales ratio of 2.5 for the month of February would be appropriate. If the sales for February were planned at $12,000, how much stock should be carried on February 1?

27. The glove department's stock at the beginning of March was $67,500, with sales for the month at $15,900. What was the stock-sales ratio for March?

28. The hosiery department sells control top pantyhose at the rate of 48 dozen per week at a retail price of $4.50 per pair. Calculate the amount of stock that should be carried to achieve a turnover of 6 for the six-month season in this category.

29. A stationary store has sales of $18,000 per week and a planned turnover of 4 for a six-month period. Calculate the amount of stock that should be carried in this store.

30. A handbag department plans a stock turn at 6.0 for a twelve-month period. What figure represents the number of weeks supply needed to achieve the desired turnover?

31. A knit accessories department has an average weekly sales figure of $14,500 and a planned turnover rate of 3.0 for the six-month period. Calculate the amount of stock to be carried.

32. The hosiery department plans a 2.5 stock turn for the Spring season with the following planned sales:

February —	$120,000	May —	$160,000
March —	140,000	June —	150,000
April —	160,000	July —	120,000

Calculate:

(a.) The average monthly sales

(b.) The average inventory.

33. Calculate the basic stock figure of a men's suit department that has annual sales of $1,650,000 and an annual stock turn of 4.0.

34. For May, the gift department had planned sales of $180,000. For the Spring season, the department's planned sales were $900,000, with a planned stock turnover of 3. Determine the BOM figure for May, using the basic stock method.

35. The small leather goods department had planned the following figures for the Fall season:

Planned total sales for season: $102,000
Planned turnover for season: 2
Planned sales for November: $ 26,000

Calculate the November BOM for this department, using the basic stock method.

36. Calculate the planned gross profit, in dollars, given the following data:

Planned dollar sales (given)	$900,000
Planned dollar markdowns (given)	135,000
Estimated planned dollar shortage	18,000
Percent planned markup on stock & purchases	48.5%
Reciprocal percent planned markup	51.5%

37. Find the planned purchases for June, if:

Planned June sales	$ 72,000
Planned June markdowns	3,000
Planned stock June 1	150,600
Planned stock July 1	155,600

38. Calculate the planned July purchases at cost using the following figures:

Planned July sales	$175,000
Planned July markdowns	20,000
Planned stock July 1	250,000
Planned stock July 31	125,000
Planned markup	52%

39. Note the following figures:

Planned September sales	$18,000
Planned September markdowns	800
Planned stock September 1	49,200
Planned stock October 1	50,400
Planned markup	51.5%

Determine:

   (a.) The planned September purchases at retail.

   (b.) The planned September purchases at cost.

   (c.) The turnover for September.

40. The hosiery department has an initial markup of 54%. The planned sales for April are $10,000 and $12,000 for May. The desired BOM stock-sales ratio for April is 2 and 1.5 for May. The planned markdowns are $600.

   Calculate:

   (a.) The planned April purchases at retail.

   (b.) The planned April purchases at cost.

41. The following figures are from a shoe department:

Planned stock November 1	$147,000
Planned stock December 1	110,000
Planned November markdowns	2,000
Planned November sales	63,000
Actual stock November 1	140,000
Actual November sales	61,000

Find:

(a.) The original planned purchases.

(b.) The adjusted planned purchases.

42. The petite sportswear department had the following seasonal planned figures for October:

	Dollars	Percentages
Sales	$150,000	
Markdowns		11%
BOM Stock	136,000	
EOM Stock	108,000	
Markup		51.5%

Determine:

(a.) The planned purchases at retail.

(b.) The planned purchases at cost.

43. The children's clothing department had projected the following figures for January:

   ▶ Sales of $46,000.
   ▶ Stock-sales ratio of 3.0.
   ▶ Reductions of $13,800.
   ▶ EOM inventory of $140,000.

   What are the planned retail purchases for January?

44. A men's furnishings department had the following figures planned for December:

Planned sales	$450,000
Planned markdowns	45,000
Planned stock (December)	900,000
Planned stock (January)	600,000

   However, the following figures reflect the actual performance:

Actual sales	$430,000
Actual stock (January 1)	$620,000

   Find:

   (a.) The original planned purchases.
   (b.) The adjusted planned purchases.

II. OPEN-TO-BUY CONTROL

Merchandise control results from effective use of data that is available through the dollar planning procedure. In purchasing merchandise a buyer is guided by the timing and quantity goals established in the six-month seasonal dollar plan. To provide an even tighter control on the amount of merchandise received in a specific period and to achieve as precisely as possible the sales and stock plans, the buyer refers to a merchandising figure called OPEN-TO-BUY. This term, abbreviated OTB, denotes the amount of unspent (i.e., order limit) money that is available for purchasing merchandise that will be delivered during a given period. Usually, it is calculated on a monthly basis and indicates that the buyer has not yet spent all of the planned purchases or receipts for the period in question. It represents the difference between the planned purchases for a period and the merchandise orders already placed for that period. Unfilled orders, generally known as open orders or "on order", should be charged to those months during which delivery is expected so that the buyer is able to control and time buying activities to correlate with selling activities. The purpose of this control is to identify the deviations between actual results and planned goals and so the buyer can take corrective measures when needed.

The planned purchase figure for a particular month indicates the sum available to purchase goods during the month, but this figure does not indicate the distribution of the money throughout the month. Experienced buyers attempt to distribute purchases over the entire month to:

▶ Reorder or replace fast-selling goods.

▶ Fill in stocks so that complete stocks are offered.

▶ Compete advantageously when buying special purchases and/or new and interesting items as they become available.

▶ Test offerings of new resources.

Information that shows current developments helps assure that all the planned factors will proceed according to plan. Consequently, it is common in large stores that the controller's office periodically issues a report. The typical report (see Figure 30) contains information that covers:

▶ **Sales, including:**

Plans for the month.

This month to date.

Adjusted plan for month.

▶ **Stocks, including:**

First of month.

Receipts to date (additions).

On hand today.

Planned EOM stock.

Adjusted EOM stock.

▶ **Purchases, including:**

Plans for month.

Adjusted plan for month.

Plan for next month.

▶ **Outstanding Orders, including:**

For delivery by month.

▶ **Markdowns, including:**

Season to date at BOM in dollars.

Season to date at BOM in %
to sales.

Month to date.

▶ **Open-to-Buy, including:**

Balance for current month.

▶ **Cumulative Markup, including:**

Plan for month.

Actual to date.

CO P1, MD A, RD 4, DG 026 SUITS — **Open-to-Buy Report** — (CGMRT0) PAGE: 58 09:50 AM

THIS WEEK (JUN:3/5) #24		Last Month (MAY)	Current Month (JUN)	JUN :1/5 #22	JUN :2/5 #23	JUN :3/5 #24	JUN :4/5 #25	JUN :5/5 #26	JUL :1/4 #27	CURR. MO. (MTD)	Next Month (JUL)	2 Nxt Month (AUG)	3 Nxt Month (SEP)
$ NET SALES	LAST YEAR	0	0	0	0	0	0	0	0	0	0	0	0
	PLAN	10,000	5,000	950	1,000	1,000	1,050	1,000	0	1,950	0	8,000	20,000
	ACTUAL	3,801	0	558	0	0			0	558	0	0	0
	% PLAN: LY	0	0	0	0				0	0	0	0	0
	% ACT.: LY	0		0	0					0			
	% ACT.: PLAN	(62)		(41)	(100)	0.0	0.0	0.0	0.0	(71)			
$ STOCK/SALES	LAST YEAR			0.0	0.0	0.0	0.0	0.0					
	ACTUAL	58.55		9.4	0.0								
$ MARK-DOWNS	LAST YEAR	0	0	0	0	0	0	0	0	0	0	0	0
	PLAN	2,000	5,000	950	1,000	1,000	1,050	1,000	0	1,950	0	0	1,000
	HOME OFF.	0		0	0					0			0
	P.O.S.	0		0	0					0			0
	TOTAL	0		0	0					0			0
MARK-DOWNS/ SALES	% LAST YEAR	0	0	0	0	0	0	0	0	0	0	0	0
	% PLAN	20	100	100	100	100	100	100	0	100	0	0	35
	% ACTUAL	0		0	0					0			5
# NET SALES	LAST YEAR	0	0	0	0	0	0	0	0	0	0	0	0
	PLAN	68		10	0	0				10			
# STOCK/SALES	LAST YEAR		0.00	0.0	0.0	0.0	0.0	0.0	0.0				
	ACTUAL			9.2	0.0								
# AVERAGE SALE	LAST YEAR	0.00		0.00	0.00	0.00	0.00	0.00	0.00	0.00	0.00	0.00	0.00
	PLAN	55.90											
	ACTUAL	0.00	0.00	55.80	0.00					55.80			
# AVERAGE STOCK	LAST YEAR	0.00		0.00	0.00	0.00	0.00	0.00	0.00	0.00	0.00	0.00	0.00
	PLAN		57.05	57.05	57.07	0.00				0.00			
	ACTUAL	58.55	57.05	57.05	57.07								
	LAST YEAR	0	0	0	0	0	0	0	0	0	0	0	0
	PLAN	22,000	10,000	5,249	4,680							15,000	30,000
	ACTUAL	6,558	5,249	5,249									
$ RECEIVED NOT SHIPPED		2,568		0	0	0				0			
$ SHIPMENTS				0	0	0				0			
MARK UP	% LAST YEAR	0	0	0	0	0				0	0	0	0
	% PLAN	32	32	32						32	0	35	35
	% ACTUAL	35		0	0					0			
# RECEIVED NOT SHIPPED		48		0	0	0				0			
# SHIPMENTS				0	0					0			

	Next Month (JUL)	2 Nxt Month (AUG)	3 Nxt Month (SEP)
# OPEN-TO-RECEIVE	0	0	0
# ON ORDER			
# OPEN TO BUY	0	0	0
$ OPEN-TO-RECEIVE	15,000	23,160	16,400
$ ON ORDER	0	0	0
% MARK UP	0	0	0
$ OPEN TO BUY	15,000	23,160	16,400
$ CUM. OPEN-TO-BUY	15,000	38,160	54,560

FIGURE 30 Open-To-Buy Report

A. Calculating Open-to-Buy at the Beginning of a Month

CONCEPT:

OTB = Planned purchases for the month minus Outstanding orders to be delivered that month

PROBLEM:

A buyer has planned January sales of $60,000, with an opening January stock planned at $50,000, a closing stock of $30,000, and markdowns planned at $500. If the orders that have already been placed for January delivery amount to $10,000 at retail, what is the buyer's January open-to-buy?

SOLUTION:

Jan. sales	$60,000
+ Jan. EOM stock	+ 30,000
+ Jan. markdowns	+ 500
	= $90,500
− Jan. BOM stock	− 50,000
Jan. planned purchases	= $40,500
− Jan. on order	− 10,000
Jan. OTB	= $30,500

Retail open-to-buy for any period can be converted to a cost open-to-buy figure. For example, $30,500 retail open-to-buy is converted to a cost figure by multiplying the retail open-to-buy by 100% (minus the planned markup percent).

B. Calculating Open-to-Buy During the Month

If the buyer wishes to calculate the OTB figure at a certain time during the period, the calculations may be based either on the predetermined planned purchases or the open-to-buy figure may be determined for the balance of the period using the planned closing stock figure. The problems that follow illustrate these calculations.

1. OTB for Balance of Month Based on Predetermined Planned Purchases

CONCEPT:

OTB for balance of month =

	Planned purchases for month
−	Merchandise received to date
−	On order
=	Open-to-buy

PROBLEM:

The merchandise plan shows that the planned purchases for September amount to $17,000. The store's records indicate that from September 1 to September 15 the department received $8,300 worth of new goods and there is an order of $700 for September delivery. What is the open-to-buy for the balance of the month?

SOLUTION:

Sept. planned purchases	$17,000
− Merchandise received	− 8,300
− On order	− 700
September OTB Balance	$ 8,000

2. OTB for Balance of Month Based on Planned Closing Stock

CONCEPT:

OTB for balance of month =

Pl. sales for balance of month
+ Pl. markdowns for balance of month
+ Pl. EOM stock
= Total merchandise requirements
− Actual stock-on-hand
− On order
= OTB for balance of the month

PROBLEM:

On September 15, an infants' department has a stock of $26,000 and merchandise on order amounting to $700. The planned sales for the balance of September are $8,000, with planned markdowns for the balance of the month at $500. The stock planned for September 30 is $31,800. What is the OTB for the balance of the month?

SOLUTION:

Planned sales for balance of month	$ 8,000
+ Planned markdowns for balance of month	+ 500
+ Planned EOM stock	+ 31,800
Total merchandise requirements	= 40,300
− Actual stock-on-hand	− 26,000
− On order	− 700
September OTB Balance	$ 13,600

45. What is the open-to-buy for a department that has planned purchases of $71,500 and outstanding orders of $74,000?

46. The actual stock on April 1 is $116,000 with "on order" in April amounting to $18,000. Sales are planned at $75,000, with markdowns estimated to be 3.5%, and the stock on April 30 planned at $112,000. What is the OTB for this month?

47. On March 13, stock-on-hand is $16,300, with planned sales for the balance of the month at $9,000. Merchandise on order comes to $3,000 and the planned April 1 stock is $10,000. Find the balance of the OTB for March.

48. The May portion of a six-month plan for a shop is as follows:

Planned May sales $37,500
Planned May markdowns   10%
Planned May BOM stock   66,000
Planned June BOM stock   59,000
Planned markup   49%

(a.) What are the planned May purchases at retail?

(b.) On May 1, the buyer is notified that $8,000 worth of goods (retail value) is on order. What is the open-to-buy at cost for May?

49. Find the OTB balance for December when on December 10 the stock-on-hand is $265,000 and the planned sales for the balance of the month are $171,000. Markdowns are planned at $11,000, planned inventory for December 31 is $150,000, with outstanding orders totaling $71,500.

50. A major West Coast retail store has a housewares department that operates on a planned stock turn figure of 4.0 for the six-month period August 1 to February 1. Find the number of weeks supply.

51. The annual turnover figure for a plus-size apparel department is projected at 5.0. Express the stated relationship in terms of weeks supply.

52. Average weekly sales in an automotive supply department of a major national chain is stated at $3,400,000. Turnover in this classification of merchandise is 4.0 annually. Using the weeks supply method, calculate the appropriate stock figure.

53. Turnover is planned at 2.5 for the six-month period starting February 1 through July 31. Average weekly sales for that period are $75,000. What average stock should be carried in this situation?

54. After careful analysis of economic projections from the U.S. Department of Commerce, a major department store chain set a figure of 7.5% as a realistic increase for this year's business volume over last. The previous year, the corporate volume was $3,700,000,000. What figure is projected for this year?

55. A Dallas-based specialty chain had sales of $196,000,000 last year. It plans $216,000,000 for the coming year. What percentage increase is being projected?

56. March figures for a shoe department are:

Planned sales	$510,000
Planned markdowns	6,000
Planned March 1 stock	906,000
Planned April 1 stock	630,000
Outstanding March orders	126,000
Planned markup	52%

Calculate:

(a.) The planned March purchases at retail.

(b.) The March open-to-buy at cost.

57. A junior sportswear buyer was shown the following data:

Planned sales	$22,000
BOM stock June	44,000
BOM stock July	40,000
Planned June markdown	1,000

Find:

(a.) The OTB at retail, if "on order" for June delivery is $16,000 at retail.

(b.) The OTB at cost, if the planned markup is 56%.

58. Note the following figures:

Planned sales	$ 85,000
Planned markdowns	1,000
Planned BOM stock	151,000
Planned EOM stock	105,000
Merchandise on order	21,000
Planned markup	51%

Determine:

(a.) The original planned purchases at cost.

(b.) The OTB at retail.

59. What is the OTB for a coat department with planned purchases of $56,000 and an "on order" of $19,000?

60. The stock-sales ratio in the men's shirt department has been set at 3.4 for June. Sales are planned at $34,500. What is the BOM stock for June?

61. Determine the BOM stock-sales ratio when the planned sales for April are $56,000 and the retail value of stock-on-hand for April 1 is $84,000.

62. The shoe department planned the following figures for October:

Planned sales	$46,000
Planned markdowns	6,500
Planned September EOM stock	68,000
Planned October EOM stock	54,000
Planned markup	48%

(a.) Find the planned purchases for October at retail and at cost.

(b.) On October 1, the buyer calculates that $16,000 worth of merchandise at retail is on order. What is the OTB at cost for October?

63. A boutique had net sales of $760,000 for a six-month period ending July 31. The monthly retail inventories were:

Dates	Stock-on-Hand
Feb. 1	$ 78,000
March 1	130,000
April 1	217,600
May 1	306,600
June 1	197,840
July 1	132,880
July 31	118,700
	$1,181,620

(a.) Find the turnover for the six-month period.

(b.) Calculate the stock-sales ratio for May, if sales for the month were $146,000.

64. An infants' wear buyer takes a physical inventory of stock every three months. Last year, the stock counts showed the following inventory valuations:

Dates	Stock-on-Hand
January 3	$16,390
April 2	18,412
July 1	14,473
October 2	19,670
December 31	15,880
	$84,825

What was the department's average stock for last year?

65. In a junior dress department, planned sales for the six-month period February through July were $250,000. The monthly inventories at retail for this period were:

Dates	Stock-on-Hand
February 1	$ 90,000
March 1	87,000
April 1	92,000
May 1	90,000
June 1	84,000
July 1	79,000
July 31	62,000
	$584,000

(a.) What is the planned average stock for the period?

(b.) What is the planned annual turnover based on the performance of this six-month period?

(c.) If planned sales in April were $23,000, what is the stock-sales ratio for April?

66. For the Fall season, a boutique with a 3.0 turnover has planned the following sales:

August — $100,700   November — $116,900
September — 113,500   December — 118,200
October — 150,000   January — 100,000

Using the basic stock method, find the BOM stock figures for each month.

67. Consider the following figures:

	BOM stocks	Net sales
January	$ 50,000	$ 15,000
February	46,000	18,000
March	46,000	20,000
April	46,000	14,000
May	40,000	12,000
June	40,000	16,000
July	45,000	14,000
August	45,000	14,000
September	50,000	18,000
October	60,000	18,000
November	80,000	38,000
December	44,000	16,000
January (following year)	46,000	
	$638,000	$213,000

Find:

(a.) The yearly average stock figure.
(b.) The turnover for the year.
(c.) The turnover for September.
(d.) The stock-sales ratio for June.

68. The statistical division provides an accessories buyer with the following computerized information:

Planned sales for balance of month	$162,000
Planned markdowns for balance of month	4,000
Planned EOM stock	200,000
Merchandise on order this month	64,000
Stock figure this date	214,000

Determine the balance-of-the-month OTB.

*69. Describe how a buyer may use turnover and stock-sales ratio figures in planning a future departmental operation.

*70. Are turnover and stock-sales ratio related? Explain.

*71. Does an increased turnover rate always mean that the department is functioning more effectively and profitably? Explain.

---

*For research and discussion.

FIGURE 31 Six-Month Merchandising Plan

SIX-MONTH MERCHANDISING PLAN			Department Name _____		Department No. _____				
					PLAN (This Year)	**ACTUAL (Last Year)**			
		Workroom cost							
		Cash discount %							
		Season stock turnover							
		Shortage %							
		Average Stock							
		Markdown %							

SPRING 1996		**FEB.**	**MAR.**	**APR.**	**MAY**	**JUNE**	**JULY**	**SEASON TOTAL**
SALES $	Last Year	140,000	160,000	200,000	175,000	165,000	160,000	$1,000,000
	Plan							
	Percent of Increase							
	Revised							
	Actual							
RETAIL STOCK (BOM) $	Last Year	325,000	475,000	550,000	500,000	375,000	275,000	
	Plan							
	Revised							
	Actual							
MARKDOWNS $	Last Year	20,000	20,000	30,000	35,000	45,000	50,000	200,000
	Plan (dollars)							
	Plan (percent)							
	Revised							
	Actual							
RETAIL PURCHASES	Last Year	310,000	255,000	180,000	85,000	110,000	235,000	
	Plan							
	Revised							
	Actual							
PERCENT OF INITIAL MARKON	Last Year	50.4	49.8	50.8	51.3	48.0	50.8	
	Plan							
	Revised							
	Actual							
ENDING STOCK JULY 31	Last Year	300,000						
	Plan							
	Revised							
	Actual							

Comments

Merchandise Manager _____   Buyer _____

Controller _____

▶ **Note to Students: Photocopy and enlarge this form and use it for practice problems 72 and 73.**

72. When the buyer for the junior dress department started to develop the merchandise plan for the Spring season, the following data was received. It represents the department's performance of last year. Preparing to plan the goals of the department for this year, the buyer refers to the factors of the plan located on the six-month merchandising plan, provided on p. 220, Figure 31.

> (a.) What judgmental decision must be made first? Why?
>
> (b.) What figures can be calculated from the incomplete information in the plan to either improve or repeat last year's results?

73. Under the present conditions, the buyer plans a 7% sales increase because it is realistic. On the basis of this planned sales increase, project, for the forthcoming Spring season only, the monthly sales, stocks, markdowns, and retail purchases on the six-month merchandising plan provided on p.220, Figure 31. Attach a separate sheet showing all calculations.

> (a.) Justify why the particular month was changed in the monthly sales distribution.
>
> (b.) State the method of stock planning used in calculating the BOM stocks. The ending stock is $300,000 — the same as last year.
>
> (c.) Justify change, if any, in monthly planned dollar markdowns and/or seasonal total percentage.
>
> (d.) Calculate the monthly planned retail purchases.

Open-to-Buy

It is the beginning of the third week in April. Mr. Johnston, the handbag buyer for a women's specialty store in California, studies the position of this department as he approaches the second quarter of the Spring season.

Before Mr. Johnston analyzes the actual results of the season-to-date performances, he examines the classification reports for the department to ascertain if the composition of his stock is in balance with the sales records. He finds that the leather handbag category — in which the price lines are highest — has fewer unit sales, which result in a high dollar inventory. The "fashion" fabrication of the season — sisal and straw, combined with new styling — has caused an out-of-proportion increase in the unit sales of this group when compared to the sales of previous years. This trend has surpassed even Mr. Johnston's originally high expectations of this fashion. Additionally, the weather, which has turned unusually warm for this time of year, has been a catalyst to this vigorous selling in the Spring-Summer look. On the six-month merchandising plan, the season-to-date figures for the department are:

FIGURE 32 Six-Month Merchandising Plan

Department Name _____   Department No. _____

SIX-MONTH MERCHANDISING PLAN		PLAN (This Year)	ACTUAL (Last Year)
	Workroom cost		
	Cash discount %		
	Season stock turnover	2.0	
	Shortage %		
	Average Stock	$530,000	
	Markdown %	5.0%	

SPRING 1996		FEB.	MAR.	APR.	MAY	JUNE	JULY	SEASON TOTAL
SALES $	Last Year							
	Plan	80,000	120,000	145,000	300,000	315,000	100,000	$1,060,000
	Percent of Increase							
	Revised							
	Actual	78,194	119,873					
RETAIL STOCK (BOM) $	Last Year							
	Plan	350,000	550,000	650,000	700,000	650,000	450,000	
	Revised							
	Actual	325,312	552,100	651,325				
MARKDOWNS $	Last Year							
	Plan (dollars)	3,000	5,000	5,000	8,000	16,000	16,000	53,000
	Plan (percent)							
	Revised							
	Actual	3,000	5,000					
RETAIL PURCHASES	Last Year							
	Plan							
	Revised							
	Actual							
	Outstanding Orders				$160,000	50,000		
ENDING STOCK JULY 31	Last Year							
	Plan	360,000						
	Revised							
	Actual							

Comments

Merchandise Manager _____   Buyer _____

Controller _____

As Mr. Johnston examines these figures, he considers the following facts:

- The February and March sales are almost on target for the plan, but the post-Easter sales sagged. (Mr. Johnston estimates that there will be a 10% decrease in sales from the plan during the first quarter because the sales before Easter were slightly off and have been decreasing since that time.) A 5% increase is required over planned sales for the second quarter to achieve, if not surpass, the total seasonal planned goals in relation to sales, markdowns, and turnover.

- Unless immediate action is taken, the lowered actual sales, to date, will result in a higher than planned May 1 inventory figure.

- Both the dollar amount in the inventory and the outstanding orders in the straw category are entirely too low to generate possible sales increases because the demand for this merchandise has exploded. (This requires instant correction.)

- The sales at the beginning of the third week for the month are $60,000, and the inventory figure is $795,000.

To accomplish the second quarter objectives, (i.e., to attain the originally planned figures regarding sales, markdowns, and turnover), what actions must Mr. Johnston take? What are the alternatives, if any? What adjustments do you recommend? Justify your suggestions mathematically.

UNIT VI
Invoice Mathematics — Terms of Sale

OBJECTIVES

▶ Identification and recognition of the different types of discounts, including:
   Trade discounts
   Quantity discounts
   Cash discounts
▶ Calculation of net cost.
▶ Identification and understanding of the different types of dating terms, including:
   COD
   Regular (Ordinary)
   Extra
   EOM
   ROG
   Advanced (Post-dating)
▶ Calculation of discount dates and net payments dates.
▶ Knowledge and understanding of anticipation.
▶ Identification and recognition of the different types of shipping terms.

KEY TERMS

advanced dating
anticipation
cash discount
COD dating
dating terms
discount
discount date
EOM dating
extra dating
FOB city or port of destination
FOB factory

FOB retailer's premises
list price
net cost
net payment date
net terms
regular or ordinary dating
prepaid
quantity discount
ROG dating
terms of sale
trade discount

When retail buyers select merchandise, they not only agree on the cost price, they also negotiate other factors that influence final cost. These conditions of the sale that are agreed upon when the merchandise is purchased are called TERMS OF SALE. These terms deal with discounts granted, dating for payments, transportation arrangements, and shipping charges. When the discounts granted are deducted from the billed cost, NET COST remitted to the vendor is the resultant figure. When shipping charges are to be paid by the retailer, they are added to the net cost that is due the vendor, and the final or total amount to be sent the vendor is determined. Lack of familiarity with terms of sale is a serious handicap because one of the best ways to improve profits is to lower the cost of goods assuming other factors remain constant. Consequently, any factor that will increase the ever-significant gross margin figure is essential.

▶ I. TERMS OF SALE

In this section, the various types of discounts, dating, and shipping terms involved in buying goods by the retailer will be examined because much of the secondary negotiation between vendors and retail buyers revolves around these three factors.

A. Different Types of Discounts

A discount is a deduction (expressed as a percentage) off the quoted or billed cost of the merchandise. Discounts are granted by a vendor to a purchaser for various negotiated terms. While discount practices and schedules vary from one industry to another, and from firm to firm, and even within merchandise classifications, there are three basic types of discounts. Generally, these discounts are taken in the following order: 1) trade discounts; 2) quantity discounts; and 3) cash discounts. These three types of discounts are examined in the following sections.

1. Trade Discount

TRADE DISCOUNT is a percentage or a series of percentages deducted from the LIST PRICE (i.e., the theoretical retail price recommended by the manufacturer). The price that a buyer pays for merchandise is determined by deducting a percentage (called a trade discount) from the list price. It is a means of establishing the cost price of the goods. The number and/or amount of trade discounts varies according to the classification of the purchasers, (i.e., retail stores, jobbers, other middlemen, or industrial buyers). For example, a flashlight manufacturer quotes trade discounts to general retailers at 35%, department stores at 40%, chain stores at 45%, and wholesalers at 50%.

Trade discounts are deducted regardless of when the invoice is paid. In merchandise lines that customarily offer trade discounts, the list price minus the trade discount(s) is a way of quoting the cost. In some cases, this type of discount is quoted as a single percentage, (e.g., $100 list price, less 45%), or is offered as a series of discounts (e.g., $100 list price, less 30%, less 10%, less 5%). Generally, the clothing industries do not use the trade discount approach to determine the cost of an item; usually cost prices are quoted directly, (e.g., style #332 costs $18.75 each or scarves cost $120/dozen). Trade discounts deducted from an established list price provide a vendor with a mechanism for changing cost easily. For example, when a supplier shows and describes merchandise through a catalog, the price change is done expeditiously by printing new price lists with a change in the discount. It is a device whereby the various middlemen in the channels of distribution can get larger and/or more discounts than retailers either because of the functional services they provide for the manufacturer or because of the larger quantities that they purchase. On the same item, a wholesaler might receive discounts of 40% and 10%, while the retailer may receive a discount of 40% only.

CONCEPT:

Billed cost = List price – Trade discount(s)

PROBLEM:

Trade discounts on a lawnmower that "lists" as $200 are 25%, 10%, 5%. Find the billed cost on an item quoted at $200 list, less 25%, 10%, and 5%.

SOLUTION:

List price	=	$200	
Less 25%		$200.00	$200.00
		× .25	– 50.00
		$ 50.00	$150.00 (resultant figure)
Less 10%		$150.00	$150.00
		× .10	– 15.00
		$ 15.00	$135.00 (resultant figure)
		$135.00	$135.00
Less 5%		× .05	– 6.75
		$ 6.75	$128.25 (resultant figure)
Billed Cost	=	$128.25	

Please note that even though the list price is $200, the retailer does not necessarily offer the consumer the manufacturer's suggested list price. It is common practice for the buyer to use the billed cost and the appropriate markup percent to calculate the retail price desired.

When calculating trade discounts in a series, the three arithmetic methods that may be employed are (please note that in these examples the list price of an item is $200 less 40%, less 10%):

► Direct method — each discount is deducted separately and is subtracted from the previous balance to determine the net or billed cost:

$200.00	$200.00	$150.00	$150.00
× .25	– 50.00	× .10	– 15.00
$ 50.00	$150.00	$ 15.00	$135.00

$135.00	$135.00
× .05	– 6.75
$ 6.75	$128.25 Net or billed cost

► Complement method — in which the complement of the discount is used, (e.g., 100% – 40% Discount = 60% Complement of the discount). The previous balance is used for the subsequent discounts, but it is not necessary to subtract the amount of the discount from the previous balance:

$200.00 × 75% (complement of 25%) = $150.00

$150.00 × 90% (complement of 10%) = $135.00

$135.00 × 95% (complement of 5%) = $128.25 Net or billed cost

▶ "On percentage" method — in which the complements of the discounts are multiplied to calculate the "on percent". The list price is then multiplied by the "on percent" to determine the billed cost:

"On percent" — .75 × .90 × .95 = 64.125%

$$= \$200.00 \times .64.125\%$$
$$= \$128.25 \text{ Billed cost}$$

2. Quantity Discount

QUANTITY DISCOUNT (QD) is a percentage off the billed cost given by a vendor when a stipulated quantity is purchased, and is deductible regardless of when invoices are paid. Usually the purchase of a large amount of goods is involved. While quantity discounts are not customary practice in the fashion industries, they are common in home furnishings and hard goods lines. Depending on custom and practice within individual industries, this kind of discount is offered either when the stipulated quantity is purchased, or it is given for accumulated purchases over a specified period of time. The amount of discount is based on a sliding scale (i.e., the larger the purchase, the greater the percentage of quantity discount). A quantity discount is offered as an incentive for buyers to commit to purchase large amounts of goods. This is a legal practice under specific provisions of the Robinson-Patman Act of 1936.[1] Additionally, it is the buyer's responsibility to judge the merits of savings through this quantity discount against the risks of tying up more than the normal planned amounts of open-to-buy money.

CONCEPT:

$ Quantity discount = $ Billed cost × Quantity discount %

$ Net billed cost = $ Billed cost − $ Quantity discount

PROBLEM:

A cookware manufacturer's established price schedule is:

▶ A minimum initial order of $500 receives a 1% QD.

▶ An initial order of $1,250 receives 1.5%.

▶ An initial order of $2,500 receives 2%, etc.

On an order of assorted cookware amounting to $2,000, use this discount schedule to determine:

(a.) The quantity discount.

(b.) The net cost of this order.

SOLUTION:

(a.)	Quantity discount	= $2,000 × 1.5%
		= $2,000 × .015
	Quantity Discount	= $ 30
(b.)	Billed cost	= $2,000
	− Quantity discount	− 30
	Net Cost	= $1,970

3. Cash Discount

CASH DISCOUNT, the most common kind of discount, is a stated percentage of the billed cost allowed by a vendor if payment of the invoiced amount is made within a stipulated time. The full utilization of cash discounts is another approach to decreasing the total cost of merchandise that results in an

---

[1]Federal legislation regulating wholesale pricing practices.

increased profit potential. Although any invoice or bill must be paid within some specified time, the intent of the cash discount is to offer the purchaser an incentive to make early payment. The vendor sacrifices a fraction of the cost that is due to receive payment more rapidly. Cost prices that are subject to quantity and/or trade discounts may also be subject to cash discounts. Eligibility for a cash discount is contingent on the time element only. The cash discount percentage is not only written on the purchase order, but also on the vendor's invoice, (e.g., 8/10 terms refers to an 8% discount if the invoice is paid within 10 days).

a. Calculating Net Cost When Billed Cost and Cash Discount Are Known.

CONCEPT:

Net cost = Billed cost – Cash discount

PROBLEM:

The cost of a microwave oven is $60. The cash discount earned is 6%. What is the net cost paid to the manufacturer?

SOLUTION:

Cash discount	=	$60 Billed cost × 6% Cash discount
		$60.00
		× .06
Cash Discount	=	$ 3.60
Billed cost	=	$60.00
– Cash discount		– 3.60
Net Cost	=	$56.40 (Amount to be remitted)

b. Calculating Net Cost When List Price is Quoted and Cash Discount is Given.

When the cost of goods is stated by quoting a list price with a series of trade discounts, and the buyer is also eligible for a cash discount, the amount to be paid is determined by calculating the billed cost first, and then deducting the cash discount.

CONCEPT:

Billed cost – Cash discount = Net billed cost

PROBLEM:

The list price of a lawnmower is $200 less 25%, less 10%, less 5%. There is a 2% cash discount offered for payment within ten days. What is the amount to be remitted if the cash discount is earned?

SOLUTION:

List price	$200.00	$200.00	List price
Less 25%	× .25	– 50.00	1st dollar discount
	$ 50.00	$150.00	
Less 10%	$150.00	$150.00	Intermediate price
	× .10	– 15.00	2nd dollar discount
	$ 15.00	$135.00	
Less 5%	135.00	$135.00	2nd intermediate price
	× .05	– 6.75	3rd dollar discount
	$ 6.75	$128.25	
Less 2% cash disc.	128.25	Billed cost	
	× .02		
	$ 2.57	Cash discount	
	$128.25	Billed cost	
	– 2.57	Cash discount	
Net Cost	$125.68	Amount remitted	

c. Calculating Net Cost When Quantity and Cash Discounts Must be Considered.

When the purchase is large enough to become eligible for a quantity discount, and a cash discount is offered, the amount to be paid is calculated by first deducting the amount of the quantity discount and then deducting the cash discount.

CONCEPT:

Net cost = Billed cost – Quantity disc. – Cash disc.

PROBLEM:

A cookware manufacturer has established price schedules of:

► A minimum initial order of $500 receives 1% QD.
► An initial order of $1,250 receives 1.5%.
► An initial order of $2,500 receives 2%, etc.

Additionally, there is a 3% cash discount offered for payment within ten days. Based on this discount schedule, what is the net cost for a $2,000 order of assorted cookware if it is paid within 7 days?

SOLUTION:

Quantity discount	=	$2,000 Billed cost × 1.5% QD
	=	$2,000.00
	×	.015
Quantity discount	=	$ 30.00
Cost	=	$2,000.00 Billed cost
	–	30.00 QD
Cost	=	$1,970.00
Cash discount	=	$1,970.00
	×	3% Cash discount
Cash discount	=	$ 59.10
Net cost	=	$1,970.00
	–	59.10 Cash discount
Net Cost	=	$1,910.90

d. Calculating Net Cost When a List Price is Quoted and Quantity Discounts and Cash Discounts are Allowed.

When the cost of goods for some types of merchandise is stated by quoting a list price with a series of discounts and the buyer is granted a quantity discount and is also eligible for a cash discount, the amount to be remitted to the vendor is calculated by:

► First, deducting the series of trade discounts to find the billed cost.
► Second, determining the eligibility for the quantity discounts offered and deducting the percent or dollar amount from the calculated billed cost.
► Third, subtracting the cash discount from this amount.

The reason for this order is that when merchandise is quoted with trade discounts, the cost price of the goods must be established first. Only then can the buyer determine if the amount purchased satisfies the quantity specified for the quantity discount, regardless of when the invoice is paid. After this cost price is calculated, then the cash discount is subtracted — but only if payment is made within the stipulated time. Savings realized by these discounts alone have an impact on the department's profit margin and should be thoroughly understood for what they are and how they are calculated to determine the remitted cost price.

CONCEPT:

Net cost = List price − Trade discounts − Quantity discount − Cash discounts

PROBLEM:

A silverware buyer purchases 75 pairs of sterling salt and pepper shakers directly from a manufacturer who quotes the list price of this item for $125 less 25%, 10%, and 5%. The vendor's price schedule is as follows:

▶ A minimum order of $4,000 receives a 3% quantity discount.
▶ An initial order of $4,500 receives a 4% quantity discount.
▶ An initial order of $5,000 receives a 7% quantity discount.

Additionally, there is a 2% cash discount offered for payment made within 10 days. What is the net cost of this order if it is paid in eight days?

SOLUTION:

List price: $125 × 75 pairs	= $9,375.00
− First trade discount 25%	× .25
	= $2,343.75
Total list price	= $9,375.00
− First $ discount	−2,343.75
First intermediate price	= $7,031.25
	$7,031.25
− Second trade discount 10%	× .10
	= $703.125
First intermediate price	= $7,031.25
− Second $ discount	− 703.13
Second intermediate price	= $6,328.12
	$6,328.12
− Third trade discount 5%	× .05
	= $ 316.406
Second intermediate price	= $6,328.12
− Third $ discount	− 316.41
Billed Cost	= $6,011.71
	$6,011.71
− Quantity discount 7%	× .07
	= $ 420.82
Billed cost	= $6,011.71
− Quantity discount	− 420.82
	= $5,590.89
	$5,590.89
− Cash discount 2%	× .02
Cash discount	= $ 111.82
Billed cost	= $5,590.89
− Cash discount	− 111.82
Amount Remitted	= $5,479.07

B. Net Terms

NET TERMS is the expression used to refer to a condition of sale if a cash discount is neither offered nor permitted. When an item is sold under these conditions, it is expressed as a "net" arrangement.

PROBLEM:

A buyer purchases an item costing $500 with net terms. The invoice is dated October 5 and the bill is paid within 30 days. What is the amount of the bill that was paid?

SOLUTION:

$500 billed cost with "net terms" means no cash discount is allowed, so $500 is paid.

PRACTICE PROBLEMS

1. The Robner Electronic Company lists a top-of-the-line answering machine for $175. The trade discount offered is 30%. What is the billing price?

2. A guitar manufacturer sells electric guitars at $450 less 30% and 10%. What is the net cost to the buyer?

3. Dealer A sells stereo sets for $975 with a 30% trade discount. Competitor B offers a similar model at $760 with a 15% discount. Which is the better buy? By how much?

4. The quantity discount schedule of a china importer is .5% discount on orders for $1,000, 1% discount on orders for $1,500, 1.5% discount on orders for $2,500, and a 2% discount on orders for $3,000. A retailer's order of bone china luncheon sets amounted to $2,800. No other discounts were allowed on this particular group. How much should the retailer pay the importer?

5. As an incentive, a supplier of motor bikes offers an additional 10% quantity discount on orders of more than 5 bikes. The normal trade discount offered by the supplier is 40%, 10% on the suggested list price of $250 each. How much would a retailer remit if 8 bikes are purchased?

6. What is the net cost on an $850 order if the cash discount earned is 4%?

7. Calculate the net cost on an order for 6 dozen boys' swim trunks at $84/dozen. The cash discount earned is 3%.

8. If a buyer purchases 120 coats at $52 each and earns an 8% cash discount, what amount must be paid to the manufacturer?

9. A gift department received an invoice in the amount of $8,200, which represents the manufacturer's suggested retail price. The invoice also shows that a trade discount of 45% is allowed, as well as a 2% cash discount. Determine the amount of payment to the vendor.

10. A sporting goods buyer purchased 50 individual golf clubs of one style that cost $35 each. When the merchandise was received, the invoice indicated that an additional 5% quantity discount was allowed when more than 36 clubs of any one style were purchased. Additionally, a 3% cash discount was allowed when payment was made within 10 days. What amount should have been remitted to the manufacturer if payment was made in 7 days?

11. The housewares department placed an order for two dozen food processors that have a list price of $300 each, with trade discounts of 30%, 10%, and 5%. The supplier's price schedule is:

  ▶ A minimum order of $3,000 — 1%.
  ▶ An initial order of $4,000 — 1.5%.
  ▶ An initial order of $5,000 — 2%.
  ▶ Payments within 10 days — 2% cash discount.

If this invoice was paid in nine days, what amount should have been remitted to the seller?

▶ II. DATING

DATING is an agreement between the supplier and retailer whereby a specified time period for payment of an invoice is arranged. Dating arrangements vary within any particular industry and from one industry to another. Dating usually implies a cash discount and is expressed as a single term of sale, (e.g., 2/10). This means that the buyer will deduct a 2% cash discount from the billed cost if the payment is remitted to the manufacturer on or before the stipulated 10-day period.

For example:

Industry	Common Dating Practice
Ready-to-wear	8/10 EOM
Millinery	7/10 EOM
Home furnishings	2/10, net 30

In the home furnishings example, 2/10, net 30 means that a 2% discount off the billed cost is permitted if the invoice is paid within 10 days following the date of invoice. The payment of the net amount (i.e., total amount of billed cost) is required between the 11th and the 30th day following the date of the invoice.

PROBLEM:

An invoice, dated March 1, for 10 folding chairs at a cost of $24 each, carries terms of 2/10, net 30. If the bill is paid on any day from March 1 to March 11, 2% may be deducted. How much should be paid if the invoice is paid on March 8?

SOLUTION:

Each	$ 24.00
× Quantity	× 10
Total cost	$240.00
Billed cost	$240.00
× Cash discount %	× .02
Cash discount	$ 4.80
Billed cost	$240.00
− Cash discount	− 4.80
Net Cost	= $235.20

However, if the bill is paid on or after March 12, the full amount of $240 is due, as listed in the payment schedule following:

Discount Period		Net Period		Penalty Period
	Last	First Date	Last Date	Penalty
Date of	Date For	For Net	For Net	Period
Invoice	Discount	Payment	Payment	Starts
March 1	March 11	March 12	March 31	April 1

In dating that allows a 10-day discount period, the last day for the net payment can be determined by adding another 20 days to the last discount day (i.e., date). For example:

Starting with a | March 1 | Invoice date
+ | 10 | Discount days
March 11 | **Last discount date**
+ | 20 | Net payment days
March 31 | **Last net payment date**

or

Last Net Payment Date = March 1 invoice date + 30 days = March 31

A. Different Types of Dating

There are many different types of dating used in all industries. As stated before, variations can even occur within a particular industry. Generally, the nature of the goods influences the prevalent dating practices. For example, some segments of the apparel industry offer a relatively high cash discount (e.g., 8%) to induce purchasers to take advantage of the savings inherent for early or prompt payment. This is vital in an industry composed of many small businesses that are frequently undercapitalized. From the standpoint of the purchaser (i.e., retailer) who invokes the cash discount privilege, the cost of the merchandise is considerably reduced, which has potential implications for increased profits. In the determination of when to pay an invoice, the buyer must be able to distinguish between the discount date and the net payment date. The discount date is the date by which the invoice may be paid to take advantage of the discount granted, and the net payment date is the date by which the invoice must be paid to acquire a favorable credit rating and avoid possible late penalties.

The previous explanations and problems used in the material pertaining to the time period of payment reflect the customary, accepted United States practices that are associated with the various types of invoice payment dating. However, in buyer vs. seller markets and in certain other economic conditions (e.g., a recession) the adherence to these rules can and frequently does change. For example, when economic conditions are not favorable, a retailer may be prone to disregard or possibly extend the traditional negotiated time period. Under these conditions, the vendor usually makes exceptions and concentrates on invoice payment rather than the strict observance of datings. For the purposes of the identification and the application of the various types of dating, and because in times of hardship, individual arrangements can vary widely, the traditional rather than these modified practices will be examined.

1. COD Dating

COD DATING (Cash-on-delivery) is a type of dating that means payment must be made "on the spot" as delivery takes place. Generally, COD dating is applicable to purchasers with poor or unproven credit ratings.

PROBLEM:

Goods valued at $500 cost are purchased by a retail store under COD dating terms. What amount must be remitted? When?

SOLUTION:

Invoiced amount = $500; Amount remitted = $500

When = As delivery takes place

2. Regular (or Ordinary) Dating

REGULAR or ORDINARY DATING is one of the most common kinds of dating. The discount period is calculated from the date of the invoice, which is usually the same date that the merchandise is shipped.

PROBLEM:

What payment should be made on an invoice for $500 dated November 16, carrying terms of 4/10, net 30?

SOLUTION:

If paid on or before November 26:

Billed cost	$500.00
– Cash discount ($500 × 4%)	– 20.00
Remitted	= $480.00

If paid between November 27 and December 16 (i.e., 20-day net payment period), no discount is permitted and the full $500 is remitted. The vendor reserves the right to charge carrying fees after the expiration of the net payment period. The exercising or bypassing of this option depends on individual cases, (e.g., credit history, the relationship of the vendor to the retailer, etc.).

3. Extra Dating

EXTRA DATING (written as an X) is calculated from the date of the invoice, with a specified number of extra days granted, during which the discount may be taken. Consequently, 2/10-60X means that the bill is payable in 10 days plus 60 extra days (i.e., a total of 70 days) from the date of the invoice to earn the 2% discount. The full amount is due after the expiration of the 70 days and the customary, though often unstated, 20-day additional net payment period follows.

PROBLEM:

An invoice dated March 16 has a billed cost of $1,800 and terms of 3/10-60X. Determine:

(a.) Final date for taking cash discount.
(b.) Cash discount earned if bill is paid June 14.
(c.) Amount due if paid in full on April 17.
(d.) Last date for net payment.

SOLUTION:

(a.) Final date for cash discount:

March 16 through 31	= 15 days
April (entire month)	= 30 days
May 1 through 25	= 25 days

End of Cash Discount Period = May 25 = 70 days

Cash Discount Allowable Through May 25 Only.

(b.) **None**

(c.) Eligible for cash discount:

$1,800 Billed cost
– 54 ($1,800 × 3% Cash discount)
$1,746 Amount Due

$1,746 **Amount Due if Paid in Full on April 17.**

(d.) Last date for net payment:

20 days after May 25 = June 14

Last Date for Net Payment: June 14.

4. EOM Dating

EOM DATING (End of month) means the cash discount period is computed from the end of the month in which the invoice is dated rather than from the date of the invoice itself. Thus, 8/10 EOM (invoice dated April 1) means that the time for payment is calculated from the end of April. Additionally, an 8% cash discount may be taken if the bill is paid by May 10, that is, ten days after the end of April. Again a twenty-day net period occurs from May 11 through May 31 (implied), during which the retailer may pay the bill in full.

(Please note: Traditionally, under EOM dating only, invoices dated on or after the 25th of any month are considered to be part of the next month's transactions.[2] For example, a bill with 8/10 EOM dated on August 26, is considered a September 1 bill and the discount period extends to October 10. Of course, arrangements of this kind vary.)

PROBLEM:

An invoice for $1,000 dated March 17 has terms of 8/10 EOM.

(a.) What is the last date for deducting 8% cash discount?

(b.) What amount will be due if the bill is paid on that date?

SOLUTION:

(a.) Discount date = April 10 (ten days after end of March)

April 10 is the Last Date for 8% Discount.

(b.) $1,000 Billed cost

– $\underline{\quad 80}$ ($ 1,000 × 8% Cash discount)

$ 920 Amount due

$920 Due if Paid by April 10.

5. ROG Dating

ROG DATING (Receipt of goods) is when the discount period is calculated from the date the goods are delivered to the retailer's premises, rather than from the date of the invoice. This type of dating is often requested by buyers located at a considerable distance from the market (or shipping point). These retailers typically receive bills a few days after shipment, but may not get delivery of the merchandise itself for a considerably longer time. Therefore, 5/10 ROG, for example, means that the bill must be paid within ten days after receipt of goods to earn the cash discount.

PROBLEM:

An invoice for $100 is dated April 4 and carries terms of 5/10 ROG. The goods arrive in the store on May 7.

(a.) What is the last date that the discount may be deducted?

(b.) How much should be remitted if payment is made on that date?

SOLUTION:

(a.) Date Goods Were Received + 10 days

May 7 + 10 days   = May 17

Last Day for Discount = May 17

(b.) $100 Billed cost – $5 ($100 × 5% Cash discount) = $95

$95 Amount Due on or Before May 17.

---

[2]Presently, it is common to apply this concept for invoices dated after the 25th of a month to other types of dating as well.

6. Advanced or Post Dating

ADVANCED or POST DATING (Seasonal discount) is the type of dating that indicates the invoice date is advanced so that additional time is allowed for payment to be made and, ultimately, for the cash discount to be deducted. The discount period is then calculated from this advanced date agreed upon by the buyer and the seller. Generally, this type of dating is used by manufacturers to persuade buyers to buy and/or receive goods earlier than they would normally. Also, it is requested by purchasers who are momentarily short of cash. Consequently, if a shipment was made on February 10 and the invoice date was May 1, terms of 2/10 would mean that payment was due on or before May 11 to qualify for a 2% cash discount. (Please note that on invoices with advanced or post dating, the payment of the invoice, at net, is delayed until the last day of the month in which the cash discount is earned, after which it is considered overdue.)

PROBLEM:

An invoice for merchandise shipped on August 18 is post-dated October 1 and carries terms of 3/10, net 30. When does the discount period expire?

SOLUTION:

3% May be Deducted if Payment is Made On or Before October 11. The Full Amount is Due at the End of the Customary Net Period on October 31.

B. Net Payment Dates

Net payment dates refer to that date by which an invoice must be paid. This date is expressed as n/30. It is considered overdue — and may be subject to an interest charge — if paid after the net payment period. Of course, there are variations for determining the final net date of an invoice. The net payment date is determined by the type of cash discount dating that has been agreed upon. The commonly used practices are:

▶ Regular dating — the full amount of the invoice is due exactly 30 days from the date of invoice.

▶ EOM, ROG, and extra dating — the full amount of the invoice is determined by adding 20 days from the expiration of the cash discount.

▶ Advanced or post dating — the payment of the net invoice is delayed until the last day of the month in which the cash discount is earned.

FIGURE 33 Summary and Application of Discount Dates and Last Date for Net Payment[3]

Type of Dating	Invoice date	Last date eligible for discount	Net amt. paid between dates below	Bill past due if paid on or later than
1. Regular (2/10 net 30)	11/16	11/26	11/27 through 12/16	12/17
2. Extra Dating (X) (3/10-90X)	6/7	9/15	9/16 through 10/5	10/6
3. End of Month (EOM) (8/10 EOM)	3/17	4/10	4/11 Through 4/30	5/1
4. Receipt of Goods (ROG) (5/10 ROG; Rec'd 4/16	4/4	4/26	4/27 through 5/16	5/17
5. Advanced or Post Dating (3/10 net 30 as of 10/1)	8/18	10/11	10/12 through 10/31	11/1

---

[3]Deadline dates used in this chart are based on the actual number of days in a given month.

12. Determine the final dates on which a cash discount may be taken for different terms on invoices dated May 15. Assume that the merchandise is received in the stores on June 2. The different terms are:

    (a.) 8/10, net 30.

    (b.) 2/10 EOM.

    (c.) 2/10-60X.

    (d.) Net 30.

    (e.) 3/10 ROG.

13. An electronics retailer buys several color TV sets that list for $725 each and is billed with trade discounts of 20% and 10%. Terms are net. What is the actual net cost price of each TV?

14. Your department receives an invoice dated October 10 in the amount of $6,750. How much must be paid on November 10 if terms are:

    (a.) 10/10, net 30.

    (b.) 10/10 EOM.

    (c.) Net.

---

[4]All dating problems are based on the actual number of days in each calendar month.

15. Merchandise that amounts to $650 at cost is shipped and invoiced on August 14. Terms are 4/10 EOM. Payment is made on August 26. How much should be remitted?

16. A buyer purchased 75 ginger jar lamps at a list price of $40 each. The trade discounts were 30%, 20%, and 5% with terms of 2/10, net 30. The lamps were shipped and billed on October 18 and were received October 22. The bill was paid on October 31. What amount was paid?

17. An invoice dated March 2 carries terms of 3/10-60X. When does the discount period expire? Explain.

18. A retailer has in stock a microcomputer that lists for $1,950. Trade discounts of 35% and 10% were quoted.

    (a.) What is the billed cost?

    (b.) At what price would the microcomputer sell if the retailer decided to apply a 48% markup?

19. If you were a buyer with an invoice that came to $100, which set of series trade discounts would you favor?

    (a.) 25%, 20%, and 15%

        or

    (b.) 50% and 10%

Explain your choice.

20. What is the "on percentage" if the trade discounts are 30%, 10%, and 5%?

21. Goods are invoiced and shipped on July 1 and received on July 15. Terms are 5/10 EOM and the invoice is for $5,600. Payment is made on August 10. How much should be remitted?

22. An invoice for $3,290, dated April 26th, covering merchandise received May 19th, carries terms of 8/10 EOM, ROG. It is paid on June 10.
    (a.) How much should have been remitted?
    (b.) What would have been remitted had ROG not been included? Why?

23. A sporting goods retailer received an invoice of $3,000, dated June 10, with terms of 2/10, n/30. What are the three alternatives available to the retailer for payment of this invoice?

▶ III. ANTICIPATION

ANTICIPATION is an extra discount, which is usually calculated at the prevailing prime rate of business interest and is subject to change based on economic conditions. This extra discount is often permitted by vendors when an invoice is paid prior to the end of the cash discount period and, again, is subject to change based on economic conditions. For the purpose of illustration only, this text will use a 6% annual rate, which equates to .5% per month, or a decimal equivalent of .005 per month.

The number of days of anticipation is based upon the number of days remaining between the actual date of payment of an invoice and the last date on which the cash discount could be taken. While some vendors do not permit an anticipation discount, alert retailers deduct it, unless a notation on the invoice expressly forbids it. Many retailers, when sending in a confirmed order, specify "anticipation allowed." Anticipation is taken by retailers because, in effect, the vendor has the use of the retailer's money ahead of the date arranged by the terms of the sale, and because of this the retailer is charging the vendor "interest" for its use. An anticipation deduction is taken in addition to any other discounts that may apply, and the deduction percentage customarily is combined with the regular cash discount percentage.

Currently, however, some retailers are anticipating invoices based on the number of days from the date of payment to the end of the net period. When using this policy, a bill that is paid after the cash discount period has ended can still be anticipated for the balance of the net period.

(Note: In the practice problems illustrating anticipation that follow this section, only the more traditional cash discount period method should be used. The following example, however, shows the calculations used for both the cash discount period method and the net payment period method.)

PROBLEM:

An invoice for $100 is dated December 4 and carries terms of 2/10-30X. Anticipation is permitted. If the bill is paid on December 14:

(a.) Calculate the number of days that were anticipated and the remittance due using the cash discount period method.

(b.) Determine the number of days that were anticipated and the remittance due using the net payment period method.

SOLUTION:

(a.) Using the traditional cash discount period method.
The cash discount period is 40 days. Because the bill is paid in 10 days, the anticipation period is 30 days.

Total $ amount of goods	= $100
Anticipation	= .5% for 30 days
Cash discount	= 2%
Cash discount + Anticipation	= 2% + .5% = 2.5%
Total discount	= $100 × .025 = $2.50
Net cost	= $100 − $2.50 = $97.50
Amount Remitted	= $97.50

(b.) Using the net payment period method or an alternate approach to anticipation.

The cash discount period is 40 days and the net payment period is 60 days. Because the bill is paid in 10 days, the anticipation period is 50 days.

(Note: Using the net payment period method, if this invoice was paid after 40 days, that is, after the cash discount period had elapsed, the buyer would still be allowed an anticipation discount for the 20 days remaining in the net payment period.)

Total $ amount of goods	= $100
Anticipation	= .82% for 50 days
Cash discount	= 2%
Cash disc. + anticipation combined	= 2% + .82% (.02 + .0082) = 2.82%
Total discount	= $100 × .0282 = $2.82
Net cost	= $100 − $2.82 = $97.18
Amount Remitted	= $97.18

▶ IV. SHIPPING TERMS

As with dating, shipping charges vary in different industries and in different situations. They are expressed as free-on-board (FOB) at a designated location. The place that is designated defines the point to which the vendor pays transportation charges and assumes risk of loss or damage and the legal title to the merchandise being shipped to a purchaser. Because the factors that determine the total cost of goods include inward freight charges, it is important that the retailer negotiate advantageous shipping terms as a means of reducing the total cost of goods. The buyer should apply ethical but firm pressure on merchandise resources for appropriate and favorable terms of sale and discounts. The most common arrangements are:

▶ **FOB retailer's premises.** Vendor pays transportation charges to the retailer's store or warehouse and, unless otherwise agreed upon, bears the risk of loss until goods are received by the retailer.

▶ **FOB factory.** Purchaser pays transportation charges from factory to purchaser's premises and, unless otherwise agreed upon, bears the risk of loss from the time the goods leave the factory.

▶ **FOB city or port of destination.** Vendor pays the transportation charges to a specified location in the city of destination and then the purchaser pays delivery charges from that point to the purchaser's premises. Unless otherwise agreed upon, the risk of loss passes from seller to buyer when goods arrive at the specified location in the city of destination.

▶ **Prepaid.** Vendor pays transportation (freight) charges to the retailer's store or warehouse when the merchandise is shipped from vendor's premises. The FOB agreement, made at the time of sale, ultimately determines whether the vendor or purchaser pays the freight charges.

24. Merchandise amounting to $6,800 was shipped on March 23 and was received the same day. Terms were 7/10 EOM and anticipation was permitted. The bill was paid on April 10. How much should have been remitted?

25. Goods that are invoiced on July 26 are received on August 10. Indicate the final discount date for net payment and the final date for net payment if terms are:

	Final Discount Date	Net Payment Date
(a.) 3/10 EOM	_____	_____
(b.) 2/10, net 30	_____	_____
(c.) 2/10-30X	_____	_____
(d.) 8/10 EOM as of Sept. 1	_____	_____
(e.) 8/10 EOM, ROG	_____	_____

---

[5]In the practice problems illustrating anticipation, only the cash discount period method should be used. For the purpose of illustration, use a 6% annual rate.

26. The following goods, shipped on June 7 and received on on July 2, are:

   ▶ 50 occasional tables at $75 list price.
   ▶ 200 folding chairs at $20 list price.

   Trade discounts are 40% and 5%, with terms of 5/10 ROG, and anticipation is permitted. What amount must be remitted if the bill is paid on July 12?

27. An invoice for $4,500 is dated and received on March 15, along with the merchandise it covers. Terms are 1/10 EOM, ROG, anticipation permitted. If the bill is paid on March 26, how much should be remitted?

28. An invoice is dated and received on April 6, which carries terms of 2/10, net 30. Anticipation is permitted. It is paid on April 16. What is the discount taken?

29. An invoice dated August 28 carrying terms of 4/10-90X is paid on October 7. Anticipation is permitted. If the billed amount is $875, what should be the remittance?

30. An invoice for $1,800 dated August 10 is paid on August 20. Terms are 8/10, net 30. Anticipation is not permitted. How much should the store's accounts payable department remit?

31. An invoice was received by Grandma's Closet in the amount of $2,100 with terms of 2/10-30X, n/90. If the invoice is dated April 20 and anticipation has been negotiated, what payment would be made on April 30?

32. Calculate the payment that should be made on August 26 by Tom's Swift Shop for merchandise amounting to $5,600. The invoice is dated August 16 and terms are 8/10-60X, anticipation allowed.

33. Marie's Book Shop received an invoice dated March 26 for books amounting to $2,500, with terms of 2/10 EOM, anticipation permitted. Indicate the amount to be remitted if payment is made on April 10.

34. On June 24, a buyer received a purchase that amounted to a total list price of $16,000. This merchandise had a trade discount of 37.5%. The buyer wanted to negotiate with the supplier the last date possible to take a cash discount. The invoice on this purchase was dated June 22. Determine the various cash payment dates for the buyer under the following purchase terms:

   (a.) 2/10, n/30.
   (b.) 2/10, n/30 ROG.
   (c.) 2/10, n/30 EOM.
   (d.) 2/10-60X, anticipation allowed.

35. An invoice for a shipment of luggage is dated September 27 and is paid on October 26. The trade discount is 45% and 5%, with terms of 2/10 EOM, anticipation permitted, FOB retailer's warehouse. There is a quantity discount of 1% on initial orders of $6,000 or more. What amount should be remitted to the vendor if the bill was for:

Quant.	Luggage Item	List Price
50	2 Suiter	$45
120	Overnighter	25
70	Dress Bag	30
90	Vacationer	60

36. A bill for $6,000 is dated November 8 for merchandise that is to arrive on November 25. What is the last date for taking the discount if terms are 7/10 ROG?

37. An invoice for $2,300 has terms of 2/10-30X, net 60, FOB factory, anticipation permitted. The vendor prepaid the shipping charges of $34. The invoice was dated August 29 and was paid on September 23. How much should the vendor have received?

38. The infants' and nursery department receives an invoice for merchandise with a list price value of $10,750. It is dated January 28 and is paid on February 23. Terms are 2/10 EOM, FOB store, with trade discounts of 40% and 10%. Anticipation is not forbidden. The manufacturer prepaid the freight of $125.

    (a.) What is the last date on which a discount may be taken?

    (b.) What is the last date for payment without penalty?

    (c.) What amount should be remitted under the conditions given in this problem?

39. An athleticwear buyer receives a shipment of 30 dozen pairs of 50/50 sweatpants costing $9 each. The invoice is dated April 26 and terms are 6/10 EOM, FOB factory. Anticipation is not permitted. Shipping charges of $92 are prepaid by the vendor.

    (a.) What is the last possible date that a discount may be taken?

    (b.) If the bill was paid on May 11, how much was remitted to the vendor?

40. What amount must be remitted on an invoice for $3,200 that is dated April 26 and paid on May 11? Terms are 8/10 EOM, anticipation permitted.

*41. If you were a small, independent dress manufacturer in a very competitive market, what terms of sale would you offer to your customers? Explain your reasoning.

*42. Your department receives a bill dated July 16, which is paid July 26. Which set of terms (5/10-30X or 7/10, net 30, anticipation permitted for both) would you choose as the most advantageous to your department? Explain your choice with supporting calculations.

*43. If you were an owner/buyer for a medium-sized specialty shop, would you always take advantage of the anticipation option when offered? Why or why not? Explain your answer.

---

*For research and discussion.

44. For the Fall season, a gift boutique received an invoice dated April 23. In negotiating terms of purchase with the resource, a series discount of 35%, 25%, and terms of 2/10, n/30 ROG were agreed upon. The retail value of the merchandise received in the store on April 28 was $1,800. How much should have been remitted to the resource if the bill was paid on May 6?

45. Jean's Fashion placed an order for 12 dozen silk scarves that cost $120 per dozen. The terms negotiated on this purchase were 2/10-30X, n/60 EOM, anticipation permitted. If the invoice is dated June 10, determine the payment to be made to the manufacturer on July 10.

46. On June 10, the junior sportswear buyer for a large department store placed an order with a resource who offered terms of 8/10, n/30 EOM, FOB warehouse, freight prepaid. Determine the required payment if the date of the invoice was August 10, the amount of the invoice was $8,700, and the bill was paid August 18.

Terms of Sale

Mr. Williams, the silverware buyer for a specialty store located in San Francisco, decides to review his vendor analysis report before making his forthcoming season's purchases. The report shows that among the top six resources, there are three relatively strong and three relatively weak suppliers in relation to the gross margin each generated. Mr. Williams realizes that some important aspects of the his job as buyer are negotiating trade discounts, quantity discounts, cash discounts, and dating, as well as transportation charges. Because any and/or all of these factors can increase the essential gross margin figure, he examines copies of past orders to determine the terms of sale on previous purchases. He discovers that some suppliers granted all his requests pertaining to discount and dating elements, certain vendors negotiated these factors only after an initial order was placed, and a considerable number allowed only the absolute minimum discount that prevailed in the market.

Because business conditions have been less than favorable, Mr. Williams feels that the success of the next season depends not only on his ability to select the most desirable items from his key resources, but also to negotiate with those vendors who offer the most advantageous terms of sale, to help maintain or improve the gross margin performance.

The first classification he shops in the market is sterling flatware. He had and continues to have strong sales in one well-advertised national brand that is distributed and can be bought directly from the manufacturer or from a jobber (i.e., middleman). Customarily, Mr. Williams places his order for these goods with the manufacturer because he is able to view the complete line of patterns and buy any quantity he needs. Additionally, he has developed a rapport with one of the salespeople who served him. This salesperson would rush special orders, would call Mr. Williams about special promotions, etc. Consequently, during market week, Mr. Williams visits the manufacturer's showroom, shops the entire line, gets delivery dates, and inquires about the current terms of sale, which are:

▶ Trade discounts from list price — 40%, less 25%.

▶ Quantity discounts offered — none.

▶ Cash discount — 2/10, n/30.

▶ FOB — Factory/shipping charges are running .5%.

When Mr. Williams returns to the store, he studies his open-to-buy for this category and decides to purchase:

▶ 15 sets of Pattern A — List price $175 for each 5-piece place setting, 40-piece service.

▶ 10 sets of Pattern B — List price $200 for each 5-piece place setting, 40-piece service.

▶ 8 sets of Pattern C — List price $225 for each 5-piece place setting, 40-piece service.

He is interrupted in his calculations by his assistant who informs him that the salesperson of an out-of-town jobber would like to speak with him. This particular supplier has tried for several seasons to get the store as an account. In the past, Mr. Williams had used local jobbers, often for immediate shipment of various items when his stock would become "low" on fast-moving items, but he had not done business before with this particular firm. Mr. Williams invites the salesperson into his office and during their conversation, Mr. Williams senses the salesperson's eagerness to open an account with him now. He deduces this because of all the concessions the salesperson is willing to offer the store. Because this jobber carries the same brand of silverware Mr. Williams has just seen at the manufacturer's showroom, he asks the prices, terms, etc. They are:

▶ Trade discount from list price — 40%, less 20%, less 5%.

▶ Quantity discount — an additional 1% for orders over $10,000; 1.5% for orders over $15,000; and 2% for orders over $20,000.

▶ Cash discount — 2/10-60X, anticipation allowed.

▶ FOB store.

With which resource should Mr. Williams place his order — the manufacturer, the jobber, or both? Justify your choice mathematically.

UNIT VII

▶ # Computer System and Control in Retailing

▶ Recognition of how computers are used in retailing.

▶ Identification and understanding of commonly used computer terms used in retailing.

▶ Comprehension of the computer operations required for merchandise control.

▶ Recognition and identification of the retail strategies related to computer technology.

▶ Knowledge of the advantages of using computers in merchandising.

▶ Identification and recognition of various computerized reports used in merchandising.

bar code

bar code reader

computer

data

data base

data processing

electronic data interchange (EDI)

electronic data processing (EDP)

input

monitor

output

optical character recognition (OCR)

POS scanning

programs

processor

quick response

shipping container marking (SCM)

terminal

universal product code (UPC)

wand scanners

▶ I. OVERVIEW OF COMPUTERS IN RETAILING

The evolution of data processing systems can probably be traced to the first manual computing device, the abacus, developed in China as early as 5000 B.C. The first generation of electronic computers, however, started in 1951, and at that time, the growth of multi-store operations mandated a need for prompt, accurate, and complete facts to make necessary merchandising decisions. In the late 20th Century, high tech has made its mark everywhere, with retailing being no exception. Since the 1960's, data processing systems have provided retailers with an uninterrupted flow of information. What is meant by data processing? It is a planned conversion of facts (as by computer) into a usable or storable form as it gathers, organizes, and summarizes information. This information system is now commonly associated with the computer; therefore, it is labeled Electronic Data Processing (EDP).

In more recent years, the increased use of these systems by department stores and small retail businesses, as well as by mass merchandising chains has had a major impact on the management of merchandise inventories and other areas in record-keeping. Today, because retailers locate their stores in distant geographic markets, the computer as a tool not only furnishes data that identifies problems but also facilitates and accelerates the decision-making process. Computers assist retailers not only with merchandising functions, but programs are available and can be used for:

▶ Accounts payable and accounts receivable.

▶ Customer billing.

▶ Customer profiling and mailing lists.

▶ Layaway, gift certificate, and credit memo control.

▶ General ledger and financial reporting.

▶ Personnel data and files.

▶ Payroll.

Information technology provides retailers with a way to integrate and improve merchandising, buying, customer service, store operations, and the financial management of the business. The three basic functions performed by computers in retailing are:

▶ Input.

▶ Processing.

▶ Output.

For example, if a departmental selling summary by units, classification, and price lines is desired, the retail price on the tag of each piece of merchandise sold is read or obtained by a chosen method. Through processing, this input of information will be summarized by the processing activity, and the output is then available in a printed report. (See Figure 34.)

GLOSSARY OF TERMS

Because the computer is so pervasive in stores and retail organizations, communication within the retail industry and its related areas (e.g., manufacturing), can be effective only with knowledge of the basic terminology and systems. In the following glossary of computer terms, focus is on the most common vocabulary and systems that are used in retailing. The glossary is broken down into two sections, Areas of Data Handling and Industry-Related Computer Strategies, with definitions of the terms kept within the retailing scope whenever possible. Please note that in both sections, the defined terms are in bold.

A. Areas of Data Handling

All retailers agree about the value of computer usage in their individual stores. This powerful force that processes information accurately and promptly is particularly appreciated in this volatile industry because it has resulted in much improved control and management of merchandise and finances. The benefits of the computer to a particular organization can be easily determined through analysis of sales, productivity, and profit. Additionally, management has an almost unlimited choice concerning what programs should be instituted, maintained, or eliminated. The following terms are the most commonly used in individual stores or organizations.

► **COMPUTER** is an electronic machine that performs high-speed mathematical or analytical calculations, or that gathers, **stores**, and correlates, or similarly prepares information obtained from data in accordance with a predetermined **program**. **Computer** is also a generic term used to describe many computing devices including:

Personal Computers (PC) or Microcomputers.

PROCESSORS or **CPU**'s.

Mainframe computers.

TERMINALS.

► **HARDWARE** is the machine(s) or component(s) in a **computer** system. It consists of devices that handle **input, processing, output**, and **storage**.

► **PROGRAMS** are step-by-step instructions that run the **computer**. They are also called **software** and/or **applications**.

► **DATA** is the raw material or **input** into the data processing system. **Data** is entered (as in data entry) and then **processed** into an organized, meaningful, useful form by the **computer** system.

► **PROCESSOR**, which is also known as a central processing unit (**CPU**), interprets and executes **programs** and instructions. It also communicates with the **input** and storage devices. The **processor** is the **computer**'s center of activity. Depending on what needs to be done with the input **data**, it can be classified, sorted, summarized, calculated, or **stored** for future use. This is referred to as **processing**. The operations that are common to processing are to:

COMPUTE, which describes the process when arithmetic calculations are performed.

RETRIEVE, which is used to request and receive a paper printout, or view information on a terminal.

UPDATE, which is used to describe **data** files that reflect changed or new information or **data**.

Generally, the **processor** is contained in a cabinet that also contains floppy and hard drives for data input and storage. The **monitor** is a separate device that indicates the status of **processing** occurring in the **computer**. A **monitor** can be either monochrome (e.g., black and white, green and black, etc.) or color, but nearly always looks like a small television set.

▶ **DATA BASE** is a collection of **data** that is organized especially for rapid retrieval. A common form of **data base** in retailing is a customer **data base**, which would show such information as:

Customer name and address.

Customer age or market group.

Customer credit history.

Customer buying history.

▶ **STORAGE** holds **data** before it is **input** to the system and is **processed**. It also holds the **processed data** that is retained for future reference and is held until it is sent to the **output** devices. Originally, computer **hardware** was all in one place, (i.e., it was centralized in one location, usually a mainframe). Although this is still common, more and more computer systems are decentralized. Though the **CPU** may be in one place, the **terminals** to access the **computer** can now be in various places within a department, a retailing establishment, or even in a different location or city.

▶ **INPUT UNITS** take **data** that is given by the computer operator (i.e., user) and send it to the **processor**. This is a primary function of a retailing **computer** and involves the capturing or obtaining of **data**. Retailing **data** may be **input** in various ways and through diverse devices, including:

TERMINALS, which have electric typewriter-type keyboards (for **inputing** the data) and television screen-type **monitors** (for reading the data).

OPTICAL CHARACTER RECOGNITION (OCR), which are devices that read the **data** from the original document. One of most common forms of this device in retailing is the **wand** scanning device that reads the price tag. Another very common kind of **OCR** is the **bar code reader** (used extensively in supermarkets) that scans and **inputs** the information from the bar codes (i.e., zebra-striped symbols) on price tags of merchandise. These are both commonly used at point-of-sale (**POS**). These devices significantly reduce human error in transcribing data.

▶ **OUTPUT** is raw **data** that is **processed** into information in usable form. Two common **output** devices are **terminals** (as described under **input**) and **printers** that produce printed reports, which have been **processed** by the **program** in use.

B. Industry-Related Computer Strategies

Presently, computer technology is having a further impact on retailing by making possible a competent industry-wide pipeline of information. The following terms reflect these industry-wide changes and strategies.

▶ **QUICK RESPONSE (QR)** is an industry-wide computerized strategy, for quick and exact replenishment of fast-selling merchandise. The benefits of **QR** are both to the industry and to the individual companies. The goal is to enable retailers to provide better customer service, and, at the same time, to realize higher bottom-line profits. It requires a relationship of trust, understanding, and partnership between manufacturers, fabric or textile suppliers, and participating retailers. This partnership allows manufacturers to expedite their manufacturing processes, while providing retailers with well-timed delivery of in-demand merchandise.

► **ELECTRONIC DATA INTERCHANGE (EDI)** allows the participating manufacturers to receive a rapid and continuous flow of detailed information about what retailers are selling and what needs to be restocked. For manufacturers the emphasis is on getting the right product to the right place at the right time for the right price. The electronic exchange of purchase orders, invoices, packing slips, and other transactions between the various levels of distribution offers the retailer a significant opportunity to decrease the labor associated with preparation, entry, and interpretation of this vast amount of paper work. Additionally, because the time element is diminished, the inventory level can be substantially reduced. To these benefits can be added the increased accuracy of the information combined with the accelerated time in exchange. To facilitate the implementation of **EDI** in the retail industry, the Voluntary Interindustry Communications Standards (VICS) has defined standard formats that enable retailers and suppliers to electronically "speak the same language".

► **UNIVERSAL PRODUCT CODE (UPC)** and **POS SCANNING** are two other strategies that make use of **EDI** and **QR**. **UPC** is a 12-digit number used to track and identify each item to the lowest level of merchandise detail. The **bar code** (as described on page 262) is comprised of a 5-digit vendor number and a 5-digit merchandise number. These 10 digits are bracketed by leading and trailing control digits. The Uniform Code Council assigns an unduplicated number to each vendor and the vendor identifies and assigns each item with an individual style number. As more vendors use the **UPC** for item identification, more retailers can reduce their ticketing time and labor, and the stream of merchandise to the selling floor is further accelerated. Because the UPC provides sales data at **POS**, retailers are able to make discerning merchandising decisions on-the-spot including the quick adjustment and control of inventory composition and levels. **POS scanning** is the technology strategy that facilitates the capture and entry of sales data through **OCR** (as described on page 262) of bar codes. **POS scanners** read the **UPC bar code** quickly and accurately. This strategy also results in decreased labor and increased precision.

► **SHIPPING CONTAINER MARKING (SCM)** is the labeling of shipping containers with numeric **bar codes** to expedite the identification and processing of these containers. Vendor-supplied shipping container codes differentiate the vendor, order number, and store location by using a separate carton number for each container in a shipment. When implemented, **SCM** will enable the retailer to receive and check a shipment without actually sorting, counting, or opening the merchandise until it reaches the selling floor. This trust requires the vendors to have exceptional accuracy in packaging and ticketing the merchandise, which has been further facilitated by **UPC**.

Because **QR** produces an improved gross margin by increasing sales and minimizing markdowns, this strategy can lead to greater sales in fashion, seasonal, or basic classifications. For fashion merchandise the lead time on initial orders can be shortened allowing the retailer to astutely forecast consumer demand. For basic and/or seasonal categories, precise sales information obtained by scanning **UPC** bar codes at **POS** encourages faster replenishment of fast-selling items, and discontinuation of slow sellers. Buyers can concentrate more on merchandising activities and less on the clerical replacement of merchandise. The amount of permanent markdowns decreases as sales climb due to the sales of merchandise assortments that are desired by the consumer. The appropriate stock-sales relationship is achieved when the improved forecasting of consumer demand is accomplished through accurate and timely information. With these tools, a retailer has the capacity to respond to that demand with precision and timing. The merchandiser can give more attention to ordering what is selling and distributing it to where it is selling.

The computer as a merchandising tool, turns available information into electronic documents. The program differences that exist, by organization, in the strategies described are minimal because industry standards are set. In summary, the addition of data processed control measures by retailers can produce greater profits for the participating members because:

▶ Buyers can focus on basic primary merchandising and customer service functions because the automatic replenishment systems and EDI reduce the amount of time necessary to communicate and track purchase orders.

▶ The automatic POS capture of sales data by bar code and wand scanners, and the SKU (stock-keeping unit) level reduces the need for store merchandise counts.

▶ The need to reticket promotional merchandise (with possible shortages resulting from missing POS prices) is avoided when price-lookup functions are used with the scanning devices.

▶ SKU checking and marking are significantly reduced as the amount of vendor ticketed merchandise increases.

▶ Vendor-applied standard SCM's on cartons dramatically reduce receiving exceptions because the electronic packing slip allows pre-authorization of vendor shipments that do not conform to the original purchase order.

▶ The flow of merchandise through the distribution center is accelerated by receipt of a pre-ship notice via EDI, the vendor pre-ticketing, and the use of SCM's.

▶ The retailer's clerical costs are substantially lowered by the electronic communication of high volume business documents, (e.g., orders, confirmations, and trade invoices).

▶ The stock turns become faster not only because of the sales increases, but the frequent deliveries from vendors result in decreased inventory levels.

▶ The increased rate of stock turns improves the GMROI, a standard that is becoming prevalent as a performance measurement in department stores.

▶ The savings of cutting expenses and reducing operational costs can be passed on to the consumer by lowering prices.

These factors, combined with sales growth, will prove to be a major competitive advantage resulting in higher profit margins.

II. PERIODIC MERCHANDISING AND OPERATING INFORMATION

The preceding six units of study have explored and discussed the various factors that comprise a total merchandising operation. The elements involved in merchandising goods have been examined, revealing those elements that may contribute to, or detract from, the ultimate profit goals for a department, division, or an entire store. For the purpose of analysis, the details of the merchandising operation are vital. Periodic merchandising records furnish the statistics that are the basis for merchandising planning and control, and make corrective action possible when necessary. While these reports are not a substitute for the buyer's own knowledge and experience, the analysis and interpretation of the information supplied are aids in the buyer's decision-making.

Successful merchandisers know that effective merchandising requires departmental controls to ensure a balanced stock, reflect customer demands, and minimize the dollar investment. The benefits of a better stock assortment and optimizing dollar investment are:

▶ Greater sales results.

▶ Fewer markdowns.

▶ Increased gross margin and GMROI.

▶ Improved rate of stock turnover.

Department controls also permit management to make a profit by judging the potential of each department and buyer.

The control and supervision of the stock of merchandise has always been a soft spot in retailing. Retailers use two major merchandise control methods. One emphasizes the physical content of the stock; the other emphasizes the size or dollar value of the merchandise. This enables them to avoid being over-stocked with slow-selling merchandise or under-stocked with "the right goods at the right time." To achieve balance between sales and stocks, data must be made available for analysis and interpreted to make the proper merchandising decisions.

From a merchandiser's viewpoint some of the advantages of using computers are:

▶ Improved decision-making because great amounts of information can be processed in a short time and weekly reports can be updated daily. Speed in recording and calculating transactions permit timely and accurate decisions.

▶ Reduced time to generate consolidation reports for those executive-level groups who do not need the detailed information of the hard copy reports (as used by the buyer) but who do require for their decision-making processes an overview of vital merchandising statistics.

▶ Increased sales through the replenishment of proper classification and price lines by directing purchases that comply with customer demand.

▶ Reduced shortages through the decrease of data input errors at POS.

▶ Decreased markdowns by providing earlier sales and merchandising information to buyers, thereby allowing corrective action on slow sellers.

▶ Reduced costs of operation.

▶ Improved selection of profitable resources.

▶ Elevated rate of turnover through fast action.

- ▶ Improved protection of gross margin through fast action.
- ▶ Improved inventory control, (i.e., better stock-sales ratios).
- ▶ Accelerated conversion of data to usable reports.
- ▶ Heightened levels of GMROI through the impact of an improved gross margin, turnover, and markup percent.

Ideally, a computerized merchandising system should provide information about the physical stock that is in tandem with dollar control. Illustrations of records that furnish the desired information are examined in the next section. The reports are divided into two major groupings. The first group shows typical applications for merchandise control; the second deals with financial or dollar control. Keep in mind that each retailer must decide what information and which reports will best serve the individual store's own situation.

A. Reports Used for Merchandise Control

The purpose of these reports in this next section is to aid the buyer in the balance and control of a stock that reflects customer demand. The following illustrations show factual information available concerning merchandise classifications and individual style activity. These records help keep track of inventory. The actual format —as well as the amount of information provided — may vary, but the function of these examples is the same.

FIGURE 34 Buyer Sales By Class By Location

		Buyer Sales by Class by Location																			
RUN DATE 03/08/96 (1) WEEK ENDING 03/07/96 (2) WEEK 1 OF 5 DEPT 674		**(4) SALES IN DOLLARS**											**(5) STOCK**				**(6) ORDERS**		**(7) % TO TOTL**		
		(a) THIS WEEK–PCT			(b) MONTH TO DATE–PCT				(c) SEASON TO DATE–PCT				ON HANDS		PLAN–PCT		CUR MTH	NXT MTH	SALE-STOCK-ORDER		
		TY	LY	CHG	TY	LY	PL	CHG	TY	LY	PL	CHG	TY	LY	EOM	CHG			STD	TW	C+N
(3) CL 88 ELEC PREPARATION	TT	40.0	21.0	90	40.0	21.0		90	143.2	98.3		46	440.5	444.8		-1	302.2		16	14	26	TT
	NY	11.3	4.8	135	11.3	4.8		135	42.1	27.4		54	77.6	66.8		16	83.9		29	17	27	NY
	BR	3.1	2.2	40	3.1	2.2		40	9.1	6.5		40	33.5	18.7		80	18.9		6	7	6	BR
	FM		.9	-100		.9		-100		3.2		-100		12.1		-100						FM
	BC	3.5	1.0	237	3.5	1.0		237	10.6	5.1		108	25.3	22.6		12	14.0		7	5	4	BC
	SH	3.0	1.2	144	3.0	1.2		144	8.5	5.8		46	28.3	13.7		106	13.8		5	6	4	SH
	NM	2.8	1.9	49	2.8	1.9		49	9.2	5.5		66	24.6	20.7		19	16.5		6	5	5	NM
	GC	3.2	.7	350	3.2	.7		350	7.3	3.7		96	21.7	15.3		42	9.2		5	4	3	GC
	PG	1.6	.9	92	1.6	.9		92	5.3	5.1		4	28.1	27.8		1	11.7		3	6	3	PG
	CH	1.9	1.0	90	1.9	1.0		90	9.2	4.9		86	40.5	17.7		129	16.3		6	9	5	CH
	WP	2.9	1.8	66	2.9	1.8		66	11.6	7.4		57	26.7	28.6		-6	15.2		8	6	5	WP
	WF	1.3	1.5	-10	1.3	1.5		-10	7.0	7.6		-8	35.6	15.7		127	12.7		4	8	4	WF
	TC	.6	1.0	-39	.6	1.0		-39	3.3	4.7		-30	29.5	23.1		27	6.4		2	6	2	TC
	KP	1.6	.3	412	1.6	.3		412	4.0	2.0		101	12.8	14.8		-13	4.0		2	2	1	KP
	WG	.9	.6	57	.9	.6		57	2.7	2.9		-6	23.7	19.9		19	6.0		1	5	1	WG
	HP												-2.2	88.6		-103	45.0				14	HP
	FA	2.2	1.2	82	2.2	1.2		82	13.4	6.5		106	34.8	26.9		30	28.5		9	7	9	FA
	ES	40.0	21.0	90	40.0	21.0		90	143.2	98.3		46	440.5	432.8		2	302.2		100	100	100	ES
	CA													12.0		-100						CA
	T-C	40.0	21.0	90	40.0	21.0		90	143.2	98.3		46	440.5	432.8		2	302.2		100	100	100	T-C
	TT	40.0	21.0	90	40.0	21.0		90	143.2	98.3		46	440.5	444.8		-1	302.2		16	14	26	TT
	MD		.1	-100		.1		-100		.9		-100										MD
	RC	120.5	52.3	130	120.5	52.3		130	259.6	170.4		52										RC
(3) CL 87 COOKING ACCESSORIE	TT	22.7	23.6	-4	22.7	23.6		-4	141.2	107.5		31	306.5	501.0		-39	164.4		16	10	14	TT
	NY	5.5	6.0	-9	5.5	6.0		-9	40.5	26.9		51	41.0	103.6		-60	9.9		28	13	6	NY
	BR	2.6	2.6	-1	2.6	2.6		-1	13.8	11.3		22	17.4	32.9		-47	6.9		9	5	4	BR
	FM		.4	-100		.4		-100		2.0		-100		15.0		-100						FM
	BC	.8	.7	17	.8	.7		17	6.1	3.6		69	14.1	25.8		-45	5.6		4	4	3	BC
	SH	.9	1.5	-42	.9	1.5		-42	8.5	6.4		33	17.4	22.4		-22	5.6		5	5	3	SH
	NM	1.4	1.7	-16	1.4	1.7		-16	6.6	4.5		48	24.2	34.1		-29	2.0		4	7	1	NM
	GC	1.3	.2	447	1.3	.2		447	5.8	2.6		126	13.1	17.7		-26	2.2		4	4	1	GC
	PG	1.3	2.0	-35	1.3	2.0		-35	8.1	6.8		20	14.4	24.0		-40	3.1		5	4	1	PG
	CH	1.2	.3	283	1.2	.3		283	5.8	4.5		27	25.5	30.4		-16	3.8		4	8	2	CH
	WP	2.7	1.3	103	2.7	1.3		103	12.7	7.4		72	17.8	28.8		-38	4.3		8	5	2	WP
	WF	1.2	1.4	-13	1.2	1.4		-13	7.3	8.6		-16	19.2	26.7		-28	1.5		5	6		WF
	TC	.4	.4	-2	.4	.4		-2	7.9	6.1		30	16.5	27.6		-40	2.6		5	5	1	TC
	KP	.8	1.0	-25	.8	1.0		-25	4.1	3.6		15	10.6	17.4		-39	1.7		2	3	1	KP
	WG	.8	1.4	-39	.8	1.4		-39	3.8	5.4		-30	11.1	22.2		-50	2.0		2	3	1	WG
	HP												45.5	29.4		55	109.2			14	66	HP
	FA	1.8	2.7	-34	1.8	2.7		-34	10.4	7.9		31	18.7	24.3		-23	3.8		7	6	2	FA
	ES	22.7	23.6	-4	22.7	23.6		-4	141.2	107.5		31	306.5	482.4		-36	164.4		100	100	100	ES
	CA													18.6		-100						CA
	T-C	22.7	23.6	-4	22.7	23.6		-4	141.2	107.5		31	306.5	482.4		-36	164.4		100	100	100	T-C
	TT	22.7	23.6	-4	22.7	23.6		-4	141.2	107.5		31	306.5	501.0		-39	164.4		16	10	14	TT
	MD		.2	-100		.2		-100		.8		-100										MD
	RC	5.0	92.3	-95	5.0	92.3		-95	107.1	139.2		-23										RC
(3) CL 86 PERSONAL CARE	TT	17.2	9.2	87	17.2	9.2		87	59.0	53.6		10	437.3	605.6		-28	102.1		6	14	8	TT
	NY	11.1	4.7	136	11.1	4.7		136	25.2	27.0		-7	89.3	125.1		-29	9.6		42	20	9	NY
	BR	.8	.4	107	.8	.4		107	1.8	2.9		-38	31.2	30.4		-3	7.9		3	7	7	BR
	FM		.4	-100		.4		-100		1.4		-100		28.7		-100						FM
	BC	.5	.1	211	.5	.1		211	2.8	1.6		74	25.1	27.5		-9	6.6		4	5	6	BC
	SH	.5	.7	-35	.5	.7		-35	3.8	2.8		36	21.8	22.6		-3	7.6		6	4	7	SH
	NM	.6	.7	-9	.6	.7		-9	2.4	2.5		-5	24.0	34.1		-30	9.7		4	5	9	NM
	GC		-.1	-87		-.1		-87	2.2	.5		353	23.4	27.7		-16	6.5		3	5	6	GC
	PG	.4	.4	-11	.4	.4		-11	1.5	1.2		27	25.3	21.2		19	7.4		2	5	7	PG
	CH	.2	.1	231	.2	.1		231	1.3	1.9		-31	29.8	38.0		-22	6.7		2	6	6	CH
	WP	.6	.2	303	.6	.2		303	4.7	2.6		84	37.5	40.3		-7	7.5		7	8	7	WP
	WF	.3	.2	48	.3	.2		48	2.4	.7		226	26.4	23.7		11	5.3		4	6	5	WF
	TC	.6	.2	243	.6	.2		243	2.4	1.0		140	23.8	23.8			7.6		4	5	7	TC
	KP	.4	.4	9	.4	.4		9	2.3	3.0		-22	17.8	25.2		-29	6.7		3	4	6	KP
	WG	.3	.4	-17	.3	.4		-17	2.2	1.9		14	24.9	26.0		-4	5.1		3	5	5	WG
	HP												2.6	45.1		-94						HP
	FA	1.0	.5	90	1.0	.5		90	3.9	2.6		52	34.5	27.0		28	8.0		6	7	7	FA
	ES	17.2	9.2	87	17.2	9.2		87	59.0	53.6		10	437.3	566.4		-23	102.1		100	100	100	ES
	CA													39.1		-100						CA
	T-C	17.2	9.2	87	17.2	9.2		87	59.0	53.6		10	437.3	566.4		-23	102.1		100	100	100	T-C
	TT	17.2	9.2	87	1.72	9.2		87	59.0	53.6		10	437.3	605.6		-23	102.1		6	14	8	TT
	MD			-100				-100		.7		-100										MD
	RC		7.9	-100		7.9		-100	102.3	53.0		93										RC

This report provides sales information for each department, which is broken down by classification and store. It indicates by individual classifications for each branch, the sales, stock-on-hand, and outstanding orders. See p. 268 for explanation of annotations.

1. Explanation of BUYER SALES BY CLASS BY LOCATION REPORT (Figure 34)

This weekly report tabulates the sales by classification for each branch store. Each branch is identified by letter (e.g., NY). The annotated information on Figure 34 is numbered (1)-(7), and the following information corresponds to those numbers:

(1)   WEEK ENDING — Example: 3/7/93.

(2)   WEEK NUMBER — Example: week 1 of 5.

(3)   Classification numbers and names — Example: CL 88 — ELEC PREPARATION.

(4)   SALES IN DOLLARS — This category is divided into these sections:

(a) THIS WEEK — Sales in dollars, (thousands) for this week in each store, this year (TY) and last year (LY), and percent change, example: see CL 87, COOKING ACCESSORIES. ES did $22.7 TY (this week) vs. $23.6 LY (this week) with 4% decrease over LY for the week.

(b) MONTH-TO-DATE (MTD) — Sales in dollars for the month for TY, LY and percent change, example: see CL 87, COOKING ACCESSORIES. TT is the total figure MTD, TY is $22.7, LY is $23.6, with a −4% change.

(c) SEASON-TO-DATE (STD) — Sales in dollars for TY, LY, and % change for STD, example: see CL 87, COOKING ACCESSORIES, TT (total figure). $141.2 TY vs. $107.5 LY, which is a 31% change.

(5)   STOCK — On hand dollars TY vs. LY and % change, example: see CL 88, ELEC PREPARATION. Store TC has $29.5 stock on hand TY, with $23.1 stock on hand LY, which represents a 27% change.

(6)   ORDERS — Total dollar amounts of orders for the current month and next month, example: see CL 86, PERSONAL CARE. Store WP has current orders amounting to $7.5.

(7)   % TO TOTAL — % to total sales that particular classification represents STD, this week (TW), current month's orders plus any orders for the next month (C+N), example: see CL 88, Store NY, Stock-sales STD 29, TW 17, C+N 27.

PRACTICE PROBLEMS[1]

1. Compare TY with LY total dollar figures (TT), MTD, and the % of change over the last year for all three classifications (88, 87, 86).

2. Of the three classifications shown on this report, which classification is the best performing classification regarding total STD sales?

3. For CL 87, COOKING ACCESSORIES:

   (a.) Which store has the highest STD % of change? What is the percentage amount?

   (b.) What is the total STD sales for this classification this year?

   (c.) What store has the highest STD sales this year? What is the dollar amount of those sales?

   (d.) What store has the lowest STD sales this year? What is the dollar amount of those sales?

   (e.) What store has the greatest TW % change? What is the percentage amount?

---

[1]Use Figure 34 for all practice problems in this section.

4. For CL 88, ELEC PREPARATION:

(a.) What store has the highest on hand stock TY? What is the dollar amount of that stock? What is the percent change over LY? What is the current amount of on order dollars for this store?

(b.) In store WP, compare — TY over LY — TW % sales change to STD percent sales change.

(c.) What store contributed the second greatest percent of STD to the total sales?

(d.) What is the current stock-sales ratio for each branch?

(e.) Based on the current stock-sales ratios for each branch, which stores might require stock adjustments? In which stores would you increase the inventory? In which stores would you decrease the inventory?

(f.) Before any branch stock adjustments are completed, what other figures would you require to obtain an appropriate balance of sales and stock for the remainder of the month?

Applying the Concepts to Figure 34
Concept: Buyers use forms to track sales by class and by location.

Use Figure 34, lines WP-WG, for all 3 product classifications to answer the 6 questions that follow.

5.   (a.) Create a "total" for all three classifications and calculate the correct figures for every column, up through "STOCK-ON-HAND, PCT. CHG."

(b.) Compare current weekly sales with last year's weekly sales for the same period. If a negative percent change of 10% or more indicates sales problems, for which classifications and stores are difficulties with sales indicated?

(c.) What are some of the reasons that sales of these product classifications are lower than last year?

(d.) What merchandising strategies could the buyer use to increase sales of these product classifications in coming weeks?

6.   Many times, season-to-date figures serve as a more significant indicator of sales trends than do weekly sales. If a negative percent change indicates sales problems, for which classifications and stores are difficulties with sales indicated?

7.   Buyers cannot just examine total sales figures for all stores; individual store figures must also be studied. For which product classification does the total figure for percent change in weekly sales give the most misleading information? Why did this occur?

8.   (a.) Additionally, buyers can use reports to identify fast-selling merchandise as well as stores that are exceeding sales expectations. If fast-selling classifications are those that have a 25% or greater increase in sales, for which classifications and stores can fast sellers be identified?

    (b.) How can buyers use information about fast sellers in the development of merchandising strategies?

9.   (a.) Based on season-to-date figures, which product classification has shown the largest percent change in sales?

    (b.) Based on this information, what merchandising strategies might be changed?

10.   (a.) Based on current season-to-date sales, calculate the stock-sales ratio for Classification 87 for all five stores.

    (b.) If a stock-sales ratio of 2.5 or higher indicates problems, at which store(s) should the buyer attempt to reduce stock levels of this product classification?

    (c.) What merchandising strategies could be used to reduce stock levels?

FIGURE 35 Classification — Price Line Report

(1) CODE NO.	(2) CODE DESCRIPTION	(3) STORE	(4a) THIS WEEK ENDING 02/07	(4b) ONE WEEK AGO ENDING 01/31	(4c) 2 WEEKS AGO ENDING 01/24	(4d) LAST 4 WEEKS ENDING 01/10	(4e) SEASON TO DATE 02/07	(5) ON HAND UNITS	(6) ON ORDER UNITS	(7a) MTD RECEIVING	(7b) MTD TRANS-FER	(7c) MTD MARKDOWN $	(7c) MARKDOWN %	(7d) MTD NET SALES $	(7d) NET SALES %	(7e) INVENTORY STOCK SALES RATIO $	(7f) OPEN ORDER	(7g) INVENTORY + ON ORDER	CODE NO.
		14	2–	1	5	5	168	69											
		15		1	1	5	229	47											
	CODE 96 TOTAL UNITS			8	14	31	1310	431											
	TOTAL DOLLARS		2–	106	177	381	17285	5683											
96	JEANS																		96
	$15.01 – $20.00	01					48	51	120										
		05					34	15	84										
		08					31	13	48										
		10					13	1	48										
		14					26	24	48										
		15					25	25	60										
	CODE 96 TOTAL UNITS						177	129	408										
	TOTAL DOLLARS						3142	2310	6528										
96	JEANS																		96
	$20.01 – $29.00	01	4	1		5	5	5–											
		05	3			3	3	3–											
		08	2	1		3	3	3–											
		10	4	1		5	5	5–											
		14	2	1		3	3	3–											
		15	1	3		4	4	4–											
	CODE 96 TOTAL UNITS		16	7		23	23	23–											
	TOTAL DOLLARS		464	203		667	667	667–											
96	JEANS																		96
	PRICE LINE–ALL	01	6	3	1	11	391	199	120			1	30	28	20		19	47	
		05	3		3	8	325	66	84			1	19	8	9		13	21	
		08	2	5	1	8	259	111	48			1	12	14	24		8	21	
		10	4	2	3	12	188	14	48			1	25	1	1		8	9	
		14		2	6	9	238	100	48				7	14	41		8	21	
		15	2	4	1	11	285	76	60				7	11	32		10	20	
	CODE 96 TOTAL UNITS		17	16	15	59	1686	566	408										
	TOTAL DOLLARS		468	320	188	1099	23005	7656	6528			5	100	75	16		65	140	
96	CLASS TOTALS																		90
	PRICE LINE–ALL	01	108	154	302	884	7673	2394	1192			11	28	264	24		140	404	
		05	67	131	93	362	3475	1636	958	16		8	20	172	22		109	281	
		08	30	121	67	258	3054	1136	906			3	7	112	41		107	219	
		10	65	118	65	317	2680	1072	708	12		7	18	107	15		79	186	
		14	49	56	72	225	2609	1162	796			5	12	114	24		91	205	
		15	69	92	86	327	3610	1319	792	14		6	16	135	22		87	222	
	CLASS 90 TOTAL UNITS		388	672	685	2373	23101	8719	5352										
	TOTAL DOLLARS		3965	6019	5385	20605	239725	84680	77191	42		40	100	905	23		612	1516	
99	INVALID CODES																		99
	PRICE LINE–ALL	01		1	1	1	3	52	530	29				62–			80	18	
		05	1–	1–	2	3–	33–	84	378				13	5–	62		58	53	
		08				1–	31–	35	378								58	59	
		10		3		4	20–	53	378				24		2		58	58	
		14	1–	1–	1–	3–	84	70–	378				63	10–	25		58	48	
		15		1–		2	19–	29	378					7–			58	51	
	CODE 99 TOTAL UNITS		2–	1	2		16–	183	2420										
	TOTAL DOLLARS		62–	10–	37–	330–	4875–	3751	37204	29	638–	1–	100	84–	133		372	288	

Figure 35 includes only two price lines because of space, but when all the price lines within a department are given, it is possible because of this broad scope, to analyze an entire department's sales trend.

2. Explanation of CLASSIFICATION — PRICE LINE REPORT (Figure 35)

This report presents a complete selling history, by dollars and units of merchandise by classification. The control of planning and purchasing is improved by knowledge of the net sales, stock on hand, outstanding orders, receipts, transfers, and markdowns as well as the stock-sales ratio for each group. The annotated information on Figure 35 is numbered (1)-(7), and the information on p. 274 corresponds to those numbers.

(1) CODE NUMBER — Designates code within a classification, example: CODE 96.

(2) CODE DESCRIPTION — Description of codes separated by price lines, example: JEANS are separated as $15.01-$20.00. Totals include:

(a) BY CODE — Example: under STD for all price lines in the code both by units and dollars, 1,686 units, $23,005.

(b) BY CLASSIFICATION — Example: under STD, all codes and all price lines, both in units and dollars, 23,101 units, $239,725.

(c) BY DEPARTMENT — All classifications, codes, and prices lines both in units and dollars. (Not shown.)

(3) STORE — Designates by store, each store, example: 01.

(4) NET SALES UNITS — Permits analysis of unit sales. This category is divided into the following sections:

(a) THIS WEEK ENDING — Net units sold for the week ending indicated, example: 2/07.

(b) ONE WEEK AGO ENDING — Net units sold for the week previous to the date of this report, example 1/31.

(c) 2 WEEKS AGO ENDING — Shows the net sales for the week, two weeks previous to the date of this report, example 1/24.

(d) LAST 4 WEEKS AGO ENDING — An accumulated statistic of net units sold over the total last four weeks, example 1/10.

(e) SEASON TO DATE — A cumulative statistic of units and/or dollars from the beginning of the season's accounting period to present date, example: 2/07.

(5) ON HAND UNITS — Net units on hand as of the date of the report, example: code 96, Store 01 — 51 units.

(6) ON ORDER UNITS — Numbers that are shown are obtained from an order file of the purchase order, example: Store 01 — JEANS $15.01-$20 — 120 units.

(7) DOLLARS TO THE NEAREST HUNDREDS — Financial figures that reflect the actual value of the indicated transactions. This category is divided into the following sections:

(a) MTD RECEIVING — Net receipt to inventory, month-to-date.

(b) MTD TRANSFERS — Cumulative transfers between stores during the current month.

(c) MTD MARKDOWN — Cumulative markdowns processed for the current month in both dollars and percentages.

(d) MTD NET SALES — Cumulative dollar sales for the current month, by price line for each store, plus the percentage of the total company's business equaling 100% for each total.

(e) INVENTORY — The stock to sales ratio is calculated by stock-sales ratio = $\dfrac{\text{Current dollar inventory}}{\text{Net sales previous week}}$

This can be looked upon as "week's supply," example: 20 stock-sales ratio means if a particular price line continues to sell at the rate of last week's sales, and nothing new is received, it will sell out in 20 weeks.

(f) OPEN ORDER — Dollar value of the outstanding orders.

(g) INVENTORY + ON ORDER — Total liability calculated by adding the on hand to the on order figures.

11. What is the price line within CODE 96 that generates the major unit and dollar sales?

12. Which store has the best performance for CLASS TOTALS, CODE 90? How many units were sold STD?

13. Which two stores contributed 49% of the MTD sales in CODE 96 JEANS?

[2]Use Figure 35 for all practice problems in this section.

14. In the JEANS $15.01-$20.00 price lines, which store requires a "rush" delivery of its order so it can stay "in business" with this price line in this classification? How many units does this store have on hand?

15. Considering the INVENTORY and ON ORDER figures for 90 CLASS TOTALS — PRICE LINE-ALL, which two stores require adjustments of on order because of their stock-sales ratios in relation to their MTD net sales percentage? What adjustments should be made?

Applying the Concepts to Figure 35
Concept: Buyers use reports to track both sales and stock levels on a week-to-week basis.

Use Figure 35, starting at line "90 CLASS TOTALS-PRICE LINE ALL" through line "CLASS 90-TOTAL DOLLARS" to answer the 5 questions that follow.

16. For each store, calculate the percent increase or decrease in sales from the previous week to the current week. Which store has shown the largest increase?

17. What reasons might explain a decline in sales from one week to the next?

18. Since January 24, which stores show a continuous decline in sales of this merchandise classification?

19. Why must changes be made in merchandising strategies at these stores?

20. What changes in merchandising strategies would you suggest?

FIGURE 36 Classification — Stock Status Report

Column structure

(1) TWO MONTHS PRIOR	(a) OPENING STOCK, (b) SALES, (c) M/D, (d) RECEIPT
(1) ONE MONTH PRIOR	(a) OPENING STOCK, (b) SALES, (c) M/D, (d) RECEIPT
(2) CURRENT MONTH	OPENING STOCK, SALES, M/D, RECEIPT, ENDING STOCK / ON ORDER / OTB
(3) ONE MONTH FUTURE	OPENING STOCK, SALES, M/D, RECEIPT, ON ORDER / OTB

Row labels: WEEK (LY / TY), TO DATE, then MONTH (PLAN, REV PLAN, LAST YEAR)

82 BRUSHED DENIM JEANS   (4) DIVISION

Row	Two-Mo Open	Sales	M/D	Recpt	One-Mo Open	Sales	M/D	Recpt	Cur Open	Sales	M/D	Recpt	Ending/OO/OTB	Fut Open	Sales	M/D	Recpt	OO/OTB
TO DATE	106.0	88.4	1.8	57.6	80.1	18.6	2.4	18.7	132.1	9.3			ENDING STOCK 122.4; ES −.4	122.4				O.O. 62.6
PLAN	112.1	73.0		28.4	67.5	15.0			123.9	35.5	1.7	36.4	O.O.	159.0	55.0		60.0	
REV PLAN									90.0	26.0		95.0						
LAST YEAR		−.1			.1	.1		11.8	11.7	3.5		28.6	OTB 36.0	38.6	25.1		33.4	OTB 34.0

83 BRUSHED DENIM TOPS   DIVISION

Row	Two-Mo Open	Sales	M/D	Recpt	One-Mo Open	Sales	M/D	Recpt	Cur Open	Sales	M/D	Recpt	Ending/OO/OTB	Fut Open	Sales	M/D	Recpt	OO/OTB
TO DATE	73.6	47.3	1.7	12.6	33.9	6.1	1.1	25.1	65.5	4.5			ENDING STOCK 60.8; ES −.2	60.8				O.O. 4.9
PLAN	73.8	49.0		20.2	45.0	9.9			57.8	17.5	.4	21.3	O.O.	80.0	15.0		19.0	
REV PLAN									50.0	10.0		40.0						
LAST YEAR	−.1				−.1			2.7	2.7	1.6		9.4	OTB 19.2	10.9	13.4		14.0	OTB 32.4

TOTAL DENIM TOPS & JEANS   DIVISION

Row	Two-Mo Open	Sales	M/D	Recpt	One-Mo Open	Sales	M/D	Recpt	Cur Open	Sales	M/D	Recpt	Ending/OO/OTB	Fut Open	Sales	M/D	Recpt	OO/OTB
TO DATE	278.1	172.1	9.0	99.3	200.5	34.7	10.0	129.4	280.1	17.5			ENDING STOCK 261.7; ES −.8	261.7				O.O. 141.9
PLAN	261.8	171.4		78.6	169.0	35.0		6.8	279.9	67.5	9.5	60.4	O.O.	349.0	92.0		104.0	
REV PLAN									225.0	49.0		173.0						
LAST YEAR	50.9	18.7	2.1	5.1	35.7	9.4	1.2	34.0	51.6	18.2	.6	62.3	OTB 87.3	98.3	63.1	.3	98.2	OTB 49.4

The accuracy of this report requires a check of purchase and transfer journals as well as transit items reports. The OTB is constantly calculated and revised. See p. 280 for explanations of annotations

3. Explanation of CLASSIFICATION — STOCK STATUS REPORT (Figure 36)

The monthly comparison, by classification, in this report provides information that shows the necessary adjustments between the opening stock, sales, markdowns, and receipts between the planned and to-date figures. The annotated information on Figure 36 is numbered (1)-(4), and the following information corresponds to those numbers:

(1) TWO MONTHS PRIOR/ONE MONTH PRIOR — History of each classification, example: 82 BRUSHED DENIM TOPS. This category is divided into the following sections:

(a.) OPENING STOCK — Example: for BRUSHED DENIM JEANS TO DATE $106.0, PLAN 112.1; TWO MONTHS PRIOR/ONE MONTH PRIOR, OPENING STOCK TO DATE is $80.1, PLAN $67.5.

(b.) SALES — Example: for 82 BRUSHED DENIM JEANS, TWO MONTHS PRIOR TO DATE $88.4, PLAN $73.0; ONE MONTH PRIOR TO DATE $18.6, PLAN $15.0.

(c.) MD — Example: MARKDOWNS, TWO MONTHS PRIOR $1.8; ONE MONTH PRIOR $2.4.

(d.) RECEIPT — Example: TWO MONTHS PRIOR TO DATE $57.6; ONE MONTH PRIOR TO DATE $18.7.

(2) CURRENT MONTH — In addition to showing the present OPENING STOCK, SALES, MARKDOWNS, and RECEIPTS, this column also includes ENDING STOCK, ON ORDER, and OTB amounts to date, example: 82 BRUSHED DENIM has a $122.4 ENDING STOCK and $36.0 OTB.

(3) ONE MONTH FUTURE — No figures are reported here, but this is the provision for future projections.

(4) DIVISION — Shows OPENING STOCK of $122.4 for BRUSHED DENIM JEANS, the outstanding orders that are added and used in the calculation of the OTB figure, example:

for BRUSHED DENIM JEANS:

OPENING STOCK	122.4
+ ON ORDER	+ 62.6
	$185.0
Adjusted to PLANNED OPENING STOCK	$159.0
+ RECEIPTS	+ 60.0
	$219.0
	$219.0
	−185.0
Gives a $34.0 OTB figure:	$ 34.0

PRACTICE PROBLEMS[3]

21. During the CURRENT MONTH, which classification had the highest sales?

22. In the CURRENT MONTH, compare CLASSIFICATION 83 BRUSHED DENIM TOPS, sales TO-DATE with PLAN.

23. What are the projected sales for the BRUSHED DENIM TOPS for ONE MONTH FUTURE?

---

[3]Use Figure 36 for all practice problems in this section.

24. Compare the ENDING STOCK of CLASS 82, BRUSHED DENIM JEANS with the PLANNED ONE MONTH FUTURE OPENING STOCK. Is it higher, lower, or on target?

25. What, if any, action should the buyer take, after reviewing the situation in Question 24?

FIGURE 37 Monthly Sales and Stock Report

(2) CODE NO.	(3) CODE DESCRIPTION	(4) STORE	(5) LAST YEAR 03/93 DOLLAR SALES	(6) THIS YEAR (a) DOLLARS (i) SALES	(ii) ON HAND	(iii) ON ORDER	(b) UNITS (i) SALES	(ii) ON HAND	(iii) ON ORDER	(7) LAST YEAR SALES APRIL DOLLARS	UNITS	MAY DOLLARS	UNITS	JUNE DOLLARS	UNITS	(2) CODE NO.
91	CLASS SLIPPERS, ETC	01	1521	4944	6461	6740	1632	2507	2228	2062	767	3640	1235	2579	905	91
		05	1505	2869	4790	5502	1129	1698	1832	1292	456	2460	796	1686	617	
		08	880	1864	2632	2888	570	814	954	1017	396	1791	740	1315	523	
		10	853	1127	2529	3115	397	812	1086	754	273	1656	640	1641	658	
		14	766	1590	3098	2493	596	1063	846	803	290	1627	575	1241	479	
		15	2290	3898	7245	6214	1841	2607	2090	2616	1257	4443	1614	2884	1119	
	CODE 91 TOTAL		7815	16292	26755	26952	6165	9501	9036	8544	3439	15797	5600	11346	4301	
20	CLASS BODY SUITS	01	2652	1621	9360	4591	283	1651	1620	2700	510	3064	578	1932	365	20
		05	1206	1069	5726	2828	194	903	1200	1285	238	1711	356	1289	259	
		08	1250	697	4240	2525	137	737	960	1278	240	1516	281	886	174	
		10	1739	833	3808	3403	148	622	1152	1756	332	1530	280	907	179	
		14	1293	726	4688	2665	180	900	996	1254	240	1334	254	716	137	
		15	1406	960	7125	2828	190	1155	1092	1172	220	2096	398	1228	222	
	CODE 20 TOTAL		9546	5906	34947	18840	1132	5968	7020	9445	1780	11251	2147	6958	1336	
30	CLASS SOCKS & PEDS	01	6524	5922	22754	6482	6428	21366	7040	7190	8131	11957	14081	8023	9633	30
		05	5403	4673	13684	7726	4866	13310	8580	5936	7025	7683	9494	5465	6241	
		08	3576	3008	11222	4349	3076	10752	4780	3692	4043	4578	5312	3117	3753	
		10	3196	2781	11776	4411	2752	10944	4750	2998	3342	3902	4316	2630	3109	
		14	3015	2824	12323	3837	2113	11571	4240	3490	3963	4547	5446	2784	3389	
		15	3841	3469	13338	5035	3508	12582	5480	4204	4749	6311	7547	4030	4842	
	CODE 30 TOTAL		25555	22677	85097	31840	23438	80625	34870	27510	31253	38978	46196	26048	30967	
40	CLASS PANTY HOSE	01					1			5	2		1–		1–	40
	CODE 40 TOTAL						1			5	2		1–		1–	
42	FLAT STITCH SHEER PANTY HOSE	01	6157	5475	11462	14346	4454	8999	10140	3577	2951	6817	5262	5282	4056	42

The dollar sales and units on hand and on order are summarized monthly to help balance merchandise inventory to sales demand.

4. Explanation of MONTHLY SALES AND STOCK REPORT (Figure 37)

This report illustrates the monthly reporting of dollars and units for sales, on hand, and on order by classification for the individual branches. This is used to maintain balanced assortments for each branch. The annotated information on Figure 37 is numbered (1)-(7), and the following information corresponds to those numbers:

(1) MONTHLY ENDING REPORT — Example, 4/03.

(2) CODE NO — Example, CODE NO 91.

(3) CODE DESCRIPTION — Name of classification grouping, example: CLASS SLIPPERS, ETC.

(4) STORE — Identifies individual stores, example: 01.

(5) LAST YEAR — Dollar sales by classification for each store, example: CLASS SLIPPERS, STORE 01 — $1521 DOLLAR SALES LAST YEAR.

(6) THIS YEAR BY STORE — Example: In STORE 01, the following figures are listed as:

(a) DOLLARS by:

    (i) SALES — $4944.

    (ii) ON HAND — $6461.

    (iii) ON ORDER — $6740.

(b) UNITS by:

    (i) SALES — 1632 Units.

    (ii) ON HAND — 2507 Units.

    (iii) ON ORDER — 2228 Units.

(7) LAST YEAR SALES — Figures for the next three months by DOLLARS and UNITS, example: CODE 91 — CLASS SLIPPERS

LAST YEAR SALES

	APRIL	MAY	JUNE
DOLLARS	2062	3640	2579
UNITS	767	1235	905

26. For CLASS 91, CLASS SLIPPERS, ETC., examine the TOTAL figures, CODE 91, and compare LAST YEAR DOLLAR SALES to current sales for this group. What is the increase in percentage? What is the increase in dollar figures?

27. If this rate of sales increase continues (as in Problem #26), what will the estimated April sales for this year be?

28. Will the combined total inventory (ON HAND) plus total ON ORDER commitments be adequate to generate the sales estimated in Problem #27? Why or why not?

---

[4]Use Figure 37 for all practice problems in this section.

29. Store #01 contributes what percentage of sales to the total volume in CLASS 91, CLASS SLIPPERS, ETC.? The combined ON HAND and ON ORDER figures for this store is what percentage of the total department's ON ORDER and ON HAND amounts?

*30. What, if any, corrective measures should be taken? Should each individual store unit always have the same stock-sales ratio to the total figure?

31. The highest volume stores are stores #01 and #15. What are the respective stock-sales ratios for these stores?

---

*For research and discussion.

FIGURE 38 Fashion Style Status Report in Units

	(1) CODE NO	(2) VENDOR	(3) STYLE	(4) RETAIL	(5) STORE	(6a) LAST 3* DAYS ENDING	(6b) THIS WEEK ENDING 04/03	(6c) ONE WEEK AGO ENDING 03/27	(6d) 2 WEEKS AGO ENDING 03/20	(6e) LAST 4 WEEKS ENDING 04/03	(6f) SEASON TO DATE 04/03	(7) ON HAND	(8) ON ORDER	(9a) FIRST DATE	(9b) LAST DATE	(10) CUSTOMER RETURNS	(11) NOTES	CODE NO
	46	3878	6553	22.50	05	1				1	1	1-		000000	000000			46
	46	3878	6553	22.50	15		1			1	1	1-		000000	000000			46
(12)			**STYLE TOTAL**			1	1			2	2	2-						
	46	3878	7550	22.50	01		56	76		132	132	139	200	030593	031793			46
	46	3878	7550	22.50	05		21	64		85	85	5-	60	030193	031793			46
	46	3878	7550	22.50	08		22	19		41	41	14	60	030193	031793			46
	46	3878	7550	22.50	10		15	12		27	27	28	60	030193	031793			46
	46	3878	7550	22.50	14		7	3		10	10	45	60	030193	031793			46
	46	3878	7550	22.50	15		17	19		36	36	29	60	030193	031793			46
			STYLE TOTAL				138	193		331	331	250	500				FAST	
	46	3878	7552	22.50	01		311	304		615	615	202	800	030593	031793			46
	46	3878	7552	22.50	05		152	104		256	256	67	150	030593	031793			46
	46	3878	7552	22.50	08		60	54		114	114	86	150	030193	031793			46
	46	3878	7552	22.50	10		29	36		65	65	135	150	030193	031793			46
	46	3878	7552	22.50	14		29	24		53	53	147	150	030193	031793			46
	46	3878	7552	22.50	15		90	68		158	158	117	150	030193	031793			46
			STYLE TOTAL				671	590		1261	1261	754	1550			8	FAST	
	46	3878	7555	22.50	01			2		2	2	2-	800	000000	000000			46
			STYLE TOTAL					2		2	2	2-	1500					
	46	3878	7557	22.50	01		175	175		350	350	507	200	030593	031793			46
	46	3878	7557	22.50	05		46	30		76	76	219	50	030193	031793			46
	46	3878	7557	22.50	08		17	15		32	32	148	50	030193	031793			46
	46	3878	7557	22.50	10		9	8		17	17	163	50	030193	031793			46
	46	3878	7557	22.50	14		16	4		20	20	160	50	030193	031793			46
	46	3878	7557	22.50	15		39	31		70	70	200	50	030193	031793			46
			STYLE TOTAL				302	263		565	565	1397	450			1	FAST	
	46	3878	7575	22.50	08		1			1	1	1-		000000	000000			46
			STYLE TOTAL				1			1	1	1-						
	46	3878	7599	22.50	01		104	64		168	168	208		030593	031793			46
	46	3878	7599	22.50	05		15	10		25	25	75		030193	031793			46
	46	3878	7599	22.50	08		14	5		19	19	46		030193	031793			46
	46	3878	7599	22.50	10		8	9		17	17	48		030193	031793			46
	46	3878	7599	22.50	14		1	4		5	5	60		030193	031793			46
	46	3878	7599	22.50	15		16	4		20	20	70		030193	031793			46
			STYLE TOTAL				158	96		254	254	507				2	FAST	
	46	3878	7553	22.50	01							23		030593	011593			46
	46	3878	7553	22.50	05			2		2	2	2-		000000	000000			46
			STYLE TOTAL					2		2	2	21						
	46	3878	7556	22.50	08							1		000000	000000			46
			STYLE TOTAL									1						
	46	3878	7559	22.50	01		14	16		30	30	249		030593	022593			46
	46	3878	7559	22.50	08							3-		030593	100893			46
			STYLE TOTAL				14	16		30	30	246						
	46	3878	7875	22.50	01							30		000000	000000			46
			STYLE TOTAL									30						

Dept 20 — Fashion Style Status Report in Units — Week Ending 04/09/96

Tabulates information that specifically pinpoints what styles are selling and their location by store.

5. Explanation of FASHION STYLE STATUS REPORT (Figure 38)

This report enables the buyer to analyze the trend of each style and classification and to respond quickly to the selling history of each style and as well as obtaining stock-sales information. The annotated information on Figure 38 is numbered (1)-(12), and the information on p. 288 corresponds to those numbers.

(1) CODE NO — A particular code within the department listed, example: CODE 46 in DEPARTMENT 20.

(2) VENDOR — An assigned number given to each vendor used, example: VENDOR 3878.

(3) STYLE — A style number usually assigned by a vendor, example: STYLE 7550.

(4) RETAIL — Current selling price of a given style, example: $22.50.

(5) STORE — Store number which has the style number listed in stock at same time, example: STORE 01.

(6) NET SALES — This category is divided into the following sections:

(a) LAST 3 DAYS ENDING — Figures used for special reports that show sales for Monday, Tuesday, Wednesday.

(b) THIS WEEK ENDING — Figures that indicate by store, the net units sold for the 6 days preceding and including the date listed, example: WEEK ENDING 4/03.

(c) ONE WEEK AGO ENDING — Net units sold for the week previous to the "THIS WEEK ENDING", example, 3/27.

(d) 2 WEEKS AGO ENDING — Net units for the period two weeks ago, example: WEEK ENDING 3/20.

(e) LAST 4 WEEKS ENDING — Cumulative net unit sales for the last four weeks, example: 4/03.

(f) SEASON-TO-DATE — Total net units sold in a current season, example: STYLE 7550, STORE 01, SEASON-TO-DATE sold 132 pieces.

(7) ON HAND — Total units on hand by store as of the week ending that the report was published, example: WEEK ENDING 4/03, in STORE 01, STYLE 7550, had in stock 139 units.

(8) ON ORDER — Amount of any one style that has been ordered for any one store, for future delivery, example: 200 pieces of STYLE 7550 are designated for STORE 01 for future delivery.

(9) RECEIPTS — This category is divided into the following sections:

(a) FIRST DATE — First date when style was received in stock, example: STYLE 7550, 3/5.

(b) LAST DATE — Most recent date when style was received into stock, example: STYLE 7550, 3/17.

(10) CUSTOMER RETURNS — Total number of units returned by customers, example: STYLE 7552, 8 (returns).

(11) NOTES — Can indicate 1 and/or 2 messages: 1.) TRANSFER, which indicates cumulative net units transferred between stores. A plus indicates receipts into stock; a minus indicates transfers out of a store's stock; 2.) The terms FAST and SLOW can also appear in this column. They are set up to "flag" exceptional selling, which is either fast or slow. The criteria for fast or slow will have been predetermined by the buyer, indicated by code.

(12) STYLE TOTAL — A total for each style indicating the activity for this style in all stores, in units, example: STYLE 7550, 331 units sold SEASON TO DATE. These figures are located at the bottom of each listed style.

32. Analyze the STYLE TOTALS for the last 4 WEEKS ENDING 4/03. The greatest unit sales were for what VENDOR and STYLE numbers?

33. What are the 3 best selling STYLE NUMBERS in this classification? Compare, by STYLE NUMBER, the UNIT STYLE TOTALS for the WEEK ENDING 4/03, with the combined O.H. and O.O. of each STYLE. Which STYLE NUMBER requires immediate shipment for delivery?

34. For STYLE 7552, which 3 stores may not have adequate O.H. stock? What, if any, ON ORDER adjustments should be made?

---

[5]Use Figure 38 for all practice problems in this section.

35. Compare each STYLE's selling history to its ON ORDER. Which STYLE number requires an immediate review? Why?

36. Which STYLE number shows a decrease in sales from 3/27 to 4/03? From the information given in this report, can the cause of this decrease be determined? What action, if any, should be taken?

B. Reports Used for Dollar Control

The purpose of these reports is to provide the buyer with information on the actual performance and any variance from the planned performance. This information requires analysis for appropriate action and decisions. These reports provide the basis for dollar control of inventories.

FIGURE 39 Daily Flash Sales

Daily Flash Sales
Daily Flash Sales for Wednesday 10/14/96   Date: per 09 wk 2 day 4

*****PLAN NUMBERS REPRESENT AN END-OF-WEEK CUMMULATIVE TOTAL*****

LOC	(3) TODAY TY	LY	(4) PTD ($1000) TY	LY	PLAN	PCT	(5) STD TY	PLAN
(1) DEPT 220 BELTS								
(2) 01 NY	3680	2147	40	37	55	7.5 +	277	284
02 BR	301	403	4	3	5	7.8 +	32	22
03^ FM	0	0	0	0	0	0.0 +	0	0
04^ ST	0	0	0	0	0	0.0 +	0	0
05 BC	61	506	5	5	8	6.1 -	38	36
06 SH	216	207	5	4	5	27.5 +	31	28
07 CP	0	0	0	0	0	0.0 +	0	0
08 NM	719	243	7	5	8	36.3 +	57	46
09 GC	471	276	5	4	5	38.4 +	32	27
10 PG	402	77	2	2	2	25.6 +	13	8
11 CH	695	343	5	3	5	38.4 +	34	31
12 WP	322	320	6	5	7	1.4 +	41	42
13 WF	293	859	3	5	6	46.2 -	21	27
14 TC	185	237	3	4	5	9.6 -	22	23
16 KP	265	306	4	3	4	16.0 +	22	21
17 WG	177	170	2	2	2	12.4 +	17	14
20^ VV	0	0	0	0	0	0.0 +	0	0
21* FA	0	580	0	6	7	102.9 -	12	35
22* FR	0	0	0	0	0	0.0 +	0	0
24 FN	0	0	0	0	0	0.0 +	0	0
25* MI	50	0	3	0	6	100.0 +	25	29
TOT-CMP	7787	6094	90	83	117	8.4 +	638	609
TOT-ALL	7837	6674	94	90	130	4.5 +	675	673
PCT	17.4 +		4.5 +					

*****PLAN NUMBERS REPRESENT AN END-OF-WEEK CUMMULATIVE TOTAL*****

LOC	TODAY TY	LY	PTD ($1000) TY	LY	PLAN	PCT	STD TY	PLAN
DEPT 231 SCARVES								
01 NY	4822	4191	59	64	81	8.6 -	316	301
02 BR	122	59-	2	2	0	35.8 +	13	3
03^ FM	0	0	0	0	0	0.0 +	0	0
04^ ST	0	0	0	0	0	0.0 +	0	0
05 BC	661	488	8	6	10	24.8 +	43	33
06 SH	478	740	6	7	9	14.8 -	35	32
07 CP	0	0	0	0	0	0.0 +	0	0
08 NM	368	848	7	6	11	27.0 +	55	40
09 GC	407	155-	4	3	5	29.8 +	24	18
10 PG	56	44	1	0	0	131.2 +	4	1
11 CH	571	82	5	6	7	11.6 -	35	22
12 WP	73	602	10	6	10	72.4 +	53	38
13 WF	170	315	3	4	5	33.1 -	19	21
14 TC	85	483	3	3	5	17.3 -	18	20
16 KP	485	311	4	4	5	10.4 +	22	23
17 WG	26	465	3	3	4	5.3 -	19	15
20^ VV	0	0	0	0	0	0.0 +	0	0
21* FA	0	57	0	1	2	100.0 -	4	7
22* FR	0	0	0	0	0	0.0 +	0	0
24 FN	0	0	0	0	0	0.0 +	0	0
25* MI	349	0	4	0	0	100.0 +	23	0
TOT-CMP	8324	8356	113	113	152	0.2 +	658	567
TOT-ALL	8673	8413	117	115	154	2.5 +	685	574
PCT	3.0 +		2.5 +					

*****PLAN NUMBERS REPRESENT AN END-OF-WEEK CUMMULATIVE TOTAL*****

LOC	TODAY TY	LY	PTD ($1000) TY	LY	PLAN	PCT	STD TY	PLAN
MGM 105 DRESS ACCESSORIES								
01 NY	13709	12232	162	163	223	0.6 -	1038	1008
02 BR	1001	1376	16	15	20	3.9 +	122	97
03^ FM	0	0	0	0	0	0.0 +	0	0
04^ ST	0	0	0	0	0	0.0 +	0	0
05 BC	1324	2362	20	20	30	2.3 +	151	129
06 SH	1284	1786	23	21	27	12.9 +	152	136
07 CP	0	0	0	0	0	0.0 +	0	0
08 NM	2636	2871	29	29	44	0.7 -	231	215
09 GC	1163	687	14	13	17	13.8 +	96	80
10 PG	874	519	7	5	8	24.1 +	44	38
11 CH	2751	821	19	17	22	8.0 +	139	110
12 WP	968	1248	27	21	30	32.0 +	176	151
13 WF	1043	2045	11	17	21	33.4 -	83	100
14 TC	921	1389	12	13	17	10.3 -	88	80
16 KP	1504	1352	13	11	15	22.1 +	86	80
17 WG	616	1008	9	11	12	11.5 +	63	58
20^ VV	0	0	0	0	0	0.0 +	0	0
21* FA	0	1847	0	17	23	101.1 -	41	114
22* FR	0	0	0	0	0	0.0 +	0	0
24 FN	0	0	0	0	0	0.0 +	0	0
25* MI	634	0	13	0	16	100.0 +	102	85
TOT-CMP	29794	29696	363	355	486	2.2 +	2470	2282
TOT-ALL	30428	31543	376	372	525	0.9 +	2613	2481
PCT	3.5 -		0.9 +					

*****PLAN NUMBERS REPRESENT AN END-OF-WEEK CUMMULATIVE TOTAL*****

LOC	TODAY TY	LY	PTD ($1000) TY	LY	PLAN	PCT	STD TY	PLAN
DEPT 252 SOCKS								
01 NY	6382	6578	76	82	81	7.6 -	334	386
02 BR	368	313	3	3	0	0.3 -	22	7
03^ FM	0	0	0	0	0	0.0 +	0	0
04^ ST	0	0	0	0	0	0.0 +	0	0
05 BC	1017	1673	12	13	10	2.0 -	52	54
06 SH	754	1002	13	11	9	18.5 +	56	52
07 CP	0	0	0	0	0	0.0 +	0	0
08 NM	1348	1133	16	15	11	6.2 +	77	85
09 GC	839	1000	12	11	5	7.1 +	51	44
10 PG	38	27	1	1	0	16.0 -	6	7
11 CH	802	516	8	7	7	21.8 +	35	33
12 WP	1129	775	13	13	10	0.4 -	59	62
13 WF	399	958	6	9	5	35.7 -	25	42
14 TC	469	429	6	6	5	4.2 +	22	24
16 KP	335	772	4	6	5	26.2 -	19	31
17 WG	307	515	4	5	4	15.3 -	17	21
20^ VV	0	0	0	0	0	0.0 +	0	0
21* FA	0	296	0	3	2	100.0 -	5	21
22* FR	0	0	0	0	0	0.0 +	0	0
24 FN	0	0	0	0	0	0.0 +	0	0
25* MI	534	0	4	0	0	100.0 +	23	45
TOT-CMP	14187	15691	175	181	152	3.7 -	776	848
TOT-ALL	14721	15987	179	184	154	3.2 -	805	914
PCT	7.9 -		3.1 -					

The function of this report is to alert the merchandiser to the most current sales position to make any revisions deemed necessary. See p. 292 for explanation of annotations.

1. Explanation of DAILY FLASH SALES (Figure 39)

This daily report can be viewed by screen print. By department, the individual branch and the department's total sales can obtained and any necessary action can be taken quickly. The annotated information on Figure 39 is numbered (1)-(5), and the following information corresponds to those numbers:

(1) DEPARTMENT NUMBER & NAME (of item) — Example: DEPT 220, BELTS.

(2) LOCATION — Individual store number and its location, example: 01 NY.

(3) TODAY — TY actual sales and LY sales for the current day, example: in STORE 01, NY, in DEPT 220, BELTS TY $3680, LY $2147. TOT-ALL — DEPT 220, BELTS, TOTAL TY $7837, TOTAL $6674, PCT 17.4% + sales increase for TY over LY.

(4) PTD ($1000) — Planned to date of TY, LY, and PLAN figures. PLAN numbers represent an end-of-week cumulative total, example:

TOT-ALL, DEPT 220, BELTS

TY — $94

LY — $90

PCT — 4.5% + sales increase for TY over LY.

(5) CUMULATIVE TOTAL STD — STD of TY over PLAN, example:

TOT-ALL, DEPT 220, BELTS

TY — $675

PLAN — $673

37. What is the greatest total sales increase percent of TY over LY of the departments for the day reported?

38. What department had STD sales that were closest to PLAN?

39. Which department had total sales that are below PLAN? What is the percentage of variance between the STD TY sales and PLAN? What stores are ahead of PLAN?

_____

[6]Use Figure 39 for all practice problems in this section.

40. In the DRESS ACCESSORIES department, which store had the largest sales for the day? What was the dollar amount of these sales? What store had the lowest dollar sales?

41. Although the SCARF department enjoyed a 2.5% end-of-week sales increase, which stores (under PTD) show a decrease of TY over PLAN? What is the total sales of TY under PTD? What is the total sales LY under PTD?

FIGURE 40 Merchandise Statistics Report

MERCHANDISE STATISTICS REPORT

SPRING 19 ___   DEPT. ___

CURRENT WEEK — SALES, MARKDOWNS, STOCK, GROSS MARGIN DATA ESTIMATED

(1) WEEK ENDING	WK NO	(2a) Weekly TY	PLAN	LY	(2b) Cum %CHG	Cum TY	Cum PLAN	Cum LY	EXCL NEW STORE %CHG	(3) Mark On % Cum TY	LY	(4) MD $ TY	MD $ LY	MD % TY	MD % LY	(5) Stock TY	Stock PLAN	Stock LY	(6) Platform TY	LY	(7) Outstanding TY	LY
OPENING										38.2	37.8	3.6	6.6	5.7	10.5	1748.3	1400.0	1455.1				
FEB 05	01	62.5	68.8	62.7	-.4	62.5	68.8	62.7	-3.6	33.2	37.3	27.6	10.2	27.5	8.3	1954.2	1450.0	1498.8	74.7	118.0	394.5	126.9
FEB 12	02	60.2	55.0	60.6	-.7	122.7	123.8	123.5	-4.6	36.3	37.1	54.8	-47.1	26.2	27.7	1865.6	1450.0	1544.7	197.9	101.9	360.2	131.6
FEB 19	03	86.4	55.0	48.3	23.1	209.1	178.8	189.8	18.6	36.2	37.0	76.3	-29.5	27.4	12.6	1853.7	1450.0	1601.3	105.5	48.5	175.3	127.3
FEB 26	04	69.5	66.2	63.2	19.6	278.0	245.0	233.0	16.0	37.6	37.0	16.0			7.8	1888.1	1450.0	1598.4	125.9	77.7	106.7	153.2
PLAN 4					5.2				-.3	37.8	37.8				16.0				62.1	82.0	159.3	121.3
MAR 05	05	63.1	55.0	41.5	24.5	341.7	360.8	274.5	18.1	37.8	36.9	55.7	-1.2	16.3	-2.6	1901.9	1500.0	1816.5	13.9	81.3	81.3	127.0
MAR 12	06	52.1	55.0	43.9	26.7	393.6	355.0	316.5	17.0	37.7	37.0	82.8	54.0	16.0	10.7	1853.5	1500.0	1809.2	122.3	29.5	204.9	330.9
MAR 19	07	55.2	55.0	47.6	22.7	449.0	410.0	369.9	16.1	37.8	37.0	122.8	73.7	27.4	20.1	1888.7	1500.0	1554.5	7.3	52.2	164.9	340.2
MAR 26	08	62.4	75.0	76.9	15.6	511.5	485.8	442.8	6.9	37.1	37.1	38.8	78.0	60.1	17.6	1913.1	1500.0	1524.7	4.4	150.5	259.8	352.2
APR 02	09	48.1	63.5	68.8	9.0	557.5	546.5	511.6	2.8	37.8	37.2	33.4	91.2	80.1	17.8	1870.2	1500.0	1593.2				
PLAN 8					7.2				-.1	37.6	37.6	34.6		6.2								
APR 09	10		60.0	51.5			688.5	583.5			37.0		169.6	19.4			1600.0	1461.2	CUR ORD	215.1	-58.5	422.2
APR 16	11		48.0	40.5			656.5	610.5			37.1		114.8	18.8			1600.0	1515.5	NXT ORD	105.4	10.0	416.2
APR 23	12		85.0	44.8			721.5	654.5			36.9		144.2	22.8			1600.0	1458.0	FUT ORD	18.9	54.5	459.9
APR 30	13		85.2	92.2			604.7	746.7			36.9		173.7	23.3			1600.0	1401.0		64.5	18.9	193.2
PLAN 13					7.8				2.0	38.1	37.1	49.0		6.1	18.8							
MAY 07	14		145.1	151.4			944.8	898.1			37.1		188.9	18.8			1550.0	1281.2	26.7		234.0
MAY 14	15		75.0	71.0			1824.6	989.7			36.9		153.7	15.6			1550.0	1383.7	160.8		517.1
MAY 21	16		95.0	71.2			1119.8	1040.9			36.7		151.0	14.5			1550.0	1483.4	169.5		497.3
MAY 28	17		100.0	90.5			1219.8	1131.4			36.9		168.3	14.9			1550.0	1561.7	172.5		603.4
PLAN 17					7.8				2.0	36.2	36.2	63.6		5.2	15.2							
JUN 04	18		80.0	76.1			1299.8	1204.6			36.9		182.9	15.2			1500.0	1615.5	142.0		530.6
JUN 11	19		105.8	181.5			1465.6	1306.1			37.6		181.8	13.9			1500.0	1699.1	162.8		405.1
JUN 18	20		84.0	75.0			1489.6	1361.4			36.9		188.7	13.7			1500.0	1700.4	76.2		494.0
JUN 25	21		85.0	76.6			1572.6	1457.6			36.9		172.7	11.8			1500.0	1799.2	148.7		200.6
JLY 02	22		75.0	73.5			1647.6	1531.0			36.8		167.9	11.0			1500.0	1777.9	45.0		251.0
PLAN 22					7.6				1.8	36.4	36.4	75.0		4.6	8.8							
JLY 09	23		60.0	88.8			1707.6	1599.4			36.8		140.1	8.8			1500.0	1722.8	-16.1		266.9
JLY 16	24		55.0	47.7			1762.6	1647.0			36.4		141.3	8.6			1500.0	1622.4	-38.1		289.9
JLY 23	25		60.0	52.6			1822.6	1699.8			36.7		128.8	7.8			1500.0	1608.5	15.6		368.5
JLY 30	26		54.0	43.8			1876.6	1743.5			36.7		112.7	6.5			1500.0	1582.5	32.8		402.1
PLAN 26					7.6				1.8	36.5		84.4		4.5								

ACT SHTG % TY / EST SHTG % TY (TY) · ACT SHTG % LY / EST SHTG % LY (LY)

WEEKS SUPPLY: 22.0 · 18.0 · 20.7

COST PURCH & DISC EARNED INCL LOAD TO 8.6%

(8) STATISTICS BELOW ON FISCAL MONTH BASIS

	4 WEEKS TY	PLAN	LY	9 WEEKS PLAN	TY	LY	13 WEEKS PLAN	TY	LY	17 WEEKS PLAN	TY	LY	22 WEEKS PLAN	TY	LY	26 WEEKS PLAN	TY	LY
GROSS MARGIN MMO After SHTG %	17.5		29.7	41.6	30.5	30.6	22.6	30.9	18.8	24.1	31.6	24.1	32.2	32.4	26.5	28.7
Disc. Earned %	6.6	6.0	6.0	4.7	5.6	6.0	6.2	5.6	6.1	8.1	5.8	8.1	5.7	5.7	5.9	5.7
Alteration %	.1	.1	.1	.1	.1	.1	.1	.1	.1	.1	.1	.1	.1	.1
Gross Margin %	24.0	35.6	35.6	46.2	36.5	36.3	28.7	36.6	25.2	30.1	37.3	30.1	37.8	38.0	32.3	34.3
Gross Margin %	67.0	87.3	87.3	107.5	200.2	200.3	146.7	294.6	136.2	340.7	355.0	340.7	622.8	713.3	394.8	597.8
SELECTIVE DATA																
Cash Disc. To BOM V.P. %	.8	.8	.8	.8	.8	.8	.8	.8	.8	.8	.6	.3
Stock Turn	123.3	13.3	13.3	.2	.3	.4	.3	.2	.5	.7	1.3	1.0	1.1
Transactions	-6.2	-22.3	22.0	25.6	62.0	84.9	98.6
% Chg	24.70	19.36	19.36	-22.3	19.57	19.57	19.67	19.47	19.16
Avg Sales Chk.	8.2	8.1	8.1	27.49	7.9	7.8	7.3	7.3	7.7
Ret. % to Gr Sales	6.8	4.3	9.0	9.1	9.1	9.6
Selling Salary/																
Direct Pub	18.2	9.9	16.2	39.6	101.7	137.0	140.3
Gr. Lineage $	2.5	2.5	30.6	91.5	110.2	119.2
Rebates $	2.7	4.3	1.4	1.7	1.5	.9	9.1	1.2
Net Lineage %	4.6	-3.4	-5.1	-5.1	-5.3	-2.8	.4
Direct Mail $																

This report provides the buyer with information on the current operating status of the total department. This type of information controls the dollar stock and helps the buyer to make necessary adjustments to ensure a profitable season. See pp. 296–297 for explanation of annotations.

2. Explanation of MERCHANDISE STATISTICS REPORT (Figure 40)

The elements that affect profit are included in this report as well as the status of the size of the stock, and sales trends. It is a guide to the six-month projections and also shows the current performance. The annotated information on Figure 40 is numbered (1)-(8), and the following information corresponds to those numbers:

(1) WEEK ENDING and WEEK NO. — Example: FEB 05 is WEEK NO. 01.

(2) NET SALES (in 100's) — This category is divided into the following two sections:

(a) WEEKLY — Figures that show the sales of TY, PLAN, and LY, example:

WEEK ENDING FEB. 05

TY — $62.5

PLAN — $68.8

LY — $62.7

(b) CUMULATIVE — Cumulative sales for the period for TY, PLAN, LY, and % CHG, for existing stores, but also excluding any new store(s), example:

WEEK ENDING FEB. 12

TY —	$122.7
PLAN —	$123.8
LY —	$123.5
% CHG —	–.7%
EXCLUDE NEW STORE —	–4.6%

(3) MARK ON % CUM. — Markon of all merchandise handled, season to date before the adjustments of cash discounts for TY and LY, example:

WEEK ENDING FEB. 05

Opening: TY — 38.2%

LY —   37.8%

(4) MARKDOWNS-CUMULATIVE INCLUDING EMPLOYEE ALLOWANCE — Markdowns both in DOLLARS and the % (of change), TY and LY, example:

WEEK ENDING FEB. 05

	DOLLARS	%
TY —	$3.6	5.7%
LY —	$6.6	10.5%

(5) STOCK ON HAND — Merchandise in stock for TY, PLAN, and LY, example:

WEEK ENDING FEB. 05

TY — $1954.2

PLAN — $1450

LY — $1498

(6) PLATFORM RECEIPTS — Merchandise received TY and LY for the reported end of week, example:

WEEK ENDING FEB. 05

TY — $74.7

LY — $118.0

(7) OUTSTANDING ORDERS — Amount of open orders, example:

WEEK ENDING FEB. 05

TY — $394.5

LY — $126.9

(8) STATISTICS BELOW ON FISCAL MONTH BASIS — Comparison of such operational factors that affect ultimate profit as GROSS MARGIN, STOCK TURN, DISC. EARNED, etc., TY over LY, example:

9 WEEKS — GROSS MARGIN MMO After SHTG %

TY —   30.5%

PLAN — 30.6%

LY —   22.6%

NOTES

42. What was the opening CUMULATIVE MARK ON % for this department? What percent change was this over LY?

43. Season to date, which WEEK ENDING had the largest dollar sales TY? What was the dollar amount of those sales? Which WEEK ENDING had the best sales increase percent for the season TY over LY?

44. For both the months of February and March, were the sales TY larger or smaller than PLAN? State the percent of change for each month.

---

[7]Use Figure 40 for all practice problems in this section.

45. Season to date, which WEEK ENDING had the greatest dollar markdowns? What was the dollar amount of these markdowns?

46. For February, week-by-week, compare the STOCK ON HAND TY with PLAN and LY figures.

47. For the month of March, were the TY OUTSTANDING ORDERS week-by-week larger or smaller than those of LY, for the same period?

Applying the Concepts to Figure 40
Concept: Buyers use reports to plan future sales and stock levels.

Use Figure 40, from "WEEK ENDING DATE FEB 05 TO APR 30", columns "NET SALES-WEEKLY-TY/PLAN/LY" (2a) and columns "STOCK-ON-HAND TY/PLAN/LY" (5) to answer the 6 questions that follow.

48. Calculate the percent by which current sales in March were above or below planned sales.

49. Calculate the percent by which current sales in March were above or below last year's sales.

50. Based on this information, how would you evaluate the planned sales for April?

51. Last year, sales during the week ending April 23 were only $44.8; however, planned sales for this year are projected to be $85.0. What reasons could explain such a large forecasted increase?

52. Calculate the percent by which stock-on-hand in March is above or below plan.

53. What strategies could be used to reduce stock levels?

FIGURE 41 Open-To-Buy Report

(1) 10/10/96 (2) WEEK 01 OF 04 FALL 1996 OPEN TO BUY (3) DEPT 736 BRIDGE SUITS

1	(4)	(5) SALES (d) PRIOR 4 WKS		(c) 2 WKS AGO		(b) LAST WEEK		(a) CURRENT WEEK		(e) MTD		(f) STD		(g) OCT $	NOV $	DEC $	(6) STOCK ENDIN JULY $	ENDIN AUG $	(a) LAST WK	(b) CURRENT WK	(c) OCT $	NOV $	DEC $	(d) WKS OF STOCK (i) ON PL	(ii) ON TRD	
TY	+ − +	24.2		23.3		9.7		18.6		18.6		126.7							141.5	126.7				11	8	TY
PL	NY	28.5	−15	14.2	64	14.2	−32	14.4	29	14.4	29	110.5	15	57.8	43.1	49.9			177.3	166.4	166.4	142.1	99.3			PL
LY		26.5	−9	17.0	37	24.2	−60	22.2	−16	22.2	−16	135.2	−6	70.2	70.2	73.3			272.0	289.7	270.4	283.9	305.3	24		LY
TY	+ − −	1.8		2.2		1.7		1.1		1.1		13.1							23.4	21.0				10	6	TY
PL	BR	1.1	67	.6	263	.6	188	1.3	−17	1.3	−17	5.4	145	5.4	8.4	14.2			22.2	26.2	26.2	38.1	40.6			PL
LY		1.0	84	2.4	−9	3.5	−50	2.7	−60	2.7	−60	15.8	−17	10.7	10.7	6.0			40.1	40.1	26.3	18.9	39.3	33		LY
TY	− − +	6.5		3.6		2.1		4.4		4.4		22.4							45.9	42.5				37	12	TY
PL	BC	3.2	102	1.6	124	1.6	29	1.5	193	1.5	193	12.4	80	6.0	6.6	8.0			37.1	41.7	41.7	33.7	24.6			PL
LY		1.7	283	1.5	138	1.7	22	−1.0	−552	−1.0	−552	8.1	177	1.8	1.8	3.3			20.0	20.9	16.9	6.7	17.8	44		LY
TY	+ + −	7.2		3.6		4.0		3.8		3.8		26.1							32.7	37.4				19	8	TY
PL	SH	3.5	106	1.8	100	1.8	120	1.9	102	1.9	102	15.9	64	7.3	9.5	11.1			38.8	34.2	34.2	34.2	22.0			PL
LY		5.3	37	4.6	−23	5.2	−24	3.3	16	3.3	16	25.8	1	8.9	8.9	8.2			52.3	45.9	50.0	38.7	52.2	21		LY
TY	+ − −	5.3		4.9		4.0		3.8		3.8		33.9							109.8	111.4				58	58	TY
PL	NM	8.3	−36	4.2	17	4.2	−4	4.3	−11	4.3	−11	32.2	5	17.4	14.7	15.8			71.3	58.9	58.9	60.4	44.3			PL
LY		8.2	−35	4.4	13	7.0	−43	8.5	−55	8.5	−55	31.5	8	17.8	17.8	10.3			55.9	50.2	45.2	62.4	59.7	21		LY
TY	− − −	2.9		1.6		1.6		−.3		−.3		10.6							20.3	22.0				23	38	TY
PL	GC	2.7	7	1.3	25	1.3	24	1.6	−116	1.6	−116	10.7	−1	6.1	4.3	5.8			23.4	19.3	19.3	17.8	17.8			PL
LY		2.0	46	1.5	7	2.4	−32	3.2	−108	3.2	−108	11.3	−7	6.5	6.5	5.2			31.4	33.5	20.5	19.1	23.7	42		LY
TY	+ + −			.9		1.4		.1		.1		3.0							12.0	11.9				23	6	TY
PL	PG	.2	−90	.1	844	.1	256	.2	−56	.2	−56	1.3	141	1.0	1.5	2.5			2.9	6.4	6.4	9.4	12.5			PL
LY		1.1	−98	.3	216	.3	329	.2	−63	.2	−63	2.4	28	.8	.8	1.9			17.1	18.6	15.1	6.0	7.8	99		LY
TY	+ + −	2.7		1.1		1.7		.2		.2		8.6							18.2	21.0				38	54	TY
PL	CH	2.4	12	1.2	−4	1.2	38	1.2	−87	1.2	−87	9.1	−6	5.1	4.3	4.8			21.1	17.7	17.7	16.8	13.6			PL
LY		3.0	−9	.5	115	3.2	−49	2.0	−92	2.0	−92	11.1	−23	4.3	4.3	3.5			22.5	17.1	16.5	7.6	20.4	99		LY
TY	+ − −	2.6		2.3		1.5		.7		.7		10.0							15.0	16.0				74	13	TY
PL	WP	1.2	118	.6	280	.6	153	.7	1	.7	1	5.0	100	2.8	1.8	2.7			13.1	10.7	10.7	9.9	7.2			PL
LY		2.2	20	.5	318	2.8	−46	.9	−22	.9	−22	8.0	25	1.1	1.1	.6			15.3	16.2	16.2	8.5	10.3	41		LY
TY	+ − −	5.4		4.7		3.3		.9		.9		19.4							31.2	30.5				34	15	TY
PL	WF	3.5	54	1.8	161	1.8	81	1.8	−49	1.8	−49	14.1	38	7.2	5.3	5.8			29.1	27.3	27.3	22.3	15.2			PL
LY		2.5	118	4.1	16	4.5	−27	3.1	−70	3.1	−70	20.5	−5	8.9	8.9	7.9			28.2	27.1	24.9	28.9	41.8	19		LY
TY	+ − +	2.8		1.9		.8		2.5		2.5		13.2							16.3	16.8				16	10	TY
PL	TC	2.6	9	1.3	42	1.3	−39	1.2	109	1.2	109	10.3	28	5.1	1.3	5.8			26.7	23.0	23.0	19.3	13.3			PL
LY		.8	269	3.2	−41	.5	63	1.6	57	1.6	57	11.2	18	5.7	5.7	5.8			31.9	31.6	25.1	15.7	26.0	58		LY
TY	+ − +	2.7		6.1		2.3		2.9		2.9		22.0							29.6	26.6				15	8	TY
PL	KP	5.0	−46	2.5	146	2.5	−9	2.3	25	2.3	25	18.3	20	9.2	7.9	9.2			37.1	35.3	35.3	28.7	19.7			PL
LY		2.9	−6	5.3	17	5.3	−57	4.5	−36	4.5	−36	25.7	−14	13.0	13.0	7.7			68.5	68.4	60.1	28.7	52.1	43		LY
TY	+ − −	.8		1.9		1.7		1.1		1.1		10.2							32.2	27.1				63	36	TY
PL	WG	2.2	−65	1.1	74	1.1	53	1.1	−3	1.1	−3	8.6	19	4.4	3.8	4.5			21.7	16.1	16.1	13.4	7.6			PL
LY		3.9	−80	.7	183	1.9	−10	1.2	−8	1.2	−8	8.5	20	2.4	2.4	3.6			12.2	12.8	12.3	11.3	12.7	22		LY
TY												1.2														TY
PL	FA	1.3	−100	.7	−100	.7	−100	1.1	−100	1.1	−100	6.1	−80	4.1	4.3	6.9			11.4	18.2	18.2	22.3	22.7			PL
LY		1.4	−100	1.7	−100	3.1	−100	1.3	−100	1.3	−100	11.2	−89	3.4	3.4	4.5			24.9	25.6	23.6	10.3	23.0	30		LY
TY	+ − +	6.0		2.4		.9		1.7		1.7		21.9							43.1	43.4				47	63	TY
PL	MI	4.4	35	2.2	11	2.2	−3	1.8	−3	1.8	−3	17.2	27	7.2	6.3	9.2			38.2	33.7	33.7	26.7	18.6			PL
LY																										LY
TY																										TY
PL	RTV																			7.6				99	99	PL
LY																										LY
TY																										TY
PL																										PL
LY																										LY
TY																										TY
PL																										PL
LY																										LY
TY																										TY
PL																										PL
LY																										LY
TY																										TY
PL																										PL
LY																										LY
TY																										TY
PL																										PL
LY																										LY
TY																										TY
PL																										PL
LY																										LY
TY	+ − −	40.7		34.9		25.9		21.2		21.2		192.5							386.5	384.2				27	16	TY
PL	EB	35.9	13	18.1	93	18.1	43	19.1	11	19.1	11	143.2	34	77.0	72.4	90.2			343.5	316.8	316.8	304.0	238.4			PL
LY		34.3	19	29.0	21	38.3	−32	30.3	−30	30.3	−30	179.9	7	81.9	81.9	64.1			395.3	382.4	329.1	252.6	363.8	30		LY
TY	+ − +	65.0		58.2		35.6		39.8		39.8		319.2							528.0	511.0				22	13	TY
PL	ES	64.4	1	32.3	80	32.3	10	33.5	19	33.5	19	253.6	26	134.8	115.5	140.1			520.8	483.2	483.2	446.1	337.7			PL
LY		60.9	7	46.0	27	62.5	−43	52.5	−24	52.5	−24	315.1	1	152.1	152.1	137.4			667.3	672.0	599.5	536.5	669.1	27		LY
TY	+ − +	70.9		60.6		36.5		41.6		41.6		342.3							571.1	562.0				22	14	TY
PL	T	70.1	1	35.2	72	35.2	4	36.4	14	36.4	14	276.9	24	146.1	126.1	156.2			570.4	535.1	535.1	495.1	379.0			PL
LY		62.3	14	47.7	27	65.6	−44	53.7	−23	53.7	−23	326.3	5	155.5	155.5	141.9			692.2	697.6	623.1	546.7	692.0	27		LY

2		(7) ORDERS (a) PRIOR MOS. PAST DUE	OCT $	(b) NOV $	DEC $	(c) FUTURE $	(d) TOTAL $	(e) S & O	(8) MARKDOWNS (a) CURRENT WEEK	(b) MTD	(c) STD	(d) STD % SLS	(9) OPEN TO REC. OCT $	NOV $	DEC $	
TY	NY	37.0	76.9	27.0			103.9	230.7	8.0	8.0	27.7	21.9	−23.7	13.4	28.3	OTR
LY		17.2	117.2				117.2	406.9	.3	.3	43.3	32.0	65.1	40.4	28.3	PL REC
TY	BR	16.3	28.4	10.9			39.3	60.3	1.3	1.3	6.2	47.3	−16.6	14.2	23.6	OTR
LY		1.2	6.2				6.2	46.3	.2	.2	2.7	17.1	11.8	25.1	23.6	PL REC
TY	BC	6.6	6.6				6.6	49.1	2.4	2.4	8.1	36.1	4.2	2.2	3.4	OTR
LY								20.9	1.8	1.8	5.9	73.4	14.1	2.2	3.4	PL REC
TY	SH	16.5	22.7				22.7	60.1	1.4	1.4	4.8	18.4	−26.2	14.3	3.7	OTR
LY		.4	17.1				17.1	62.9			6.6	25.7	6.4	14.3	3.7	PL REC
TY	NM	9.7	10.2	8.5			18.7	130.1	−.1	−.1	7.2	21.3	−4.3	15.9	7.7	OTR
LY			8.0				8.0	58.2	.8	.8	5.7	18.0	11.3	24.4	7.7	PL REC
TY	GC	4.8	9.4				9.4	31.4	1.9	1.9	7.0	65.8	−8.3	6.2	9.3	OTR
LY								33.5	.2	.2	2.3	20.6	4.5	6.2	9.3	PL REC
TY	PG	4.9	7.2	8.2			15.3	27.3			.5	17.5	−2.1	−2.8	6.9	OTR
LY								18.6			.5	19.1	5.1	5.4	6.9	PL REC
TY	CH	11.6	12.1				12.1	33.1	.6	.6	3.1	36.4	−11.5	5.8	5.0	OTR
L								17.1	.3	.3	4.8	42.7	4.1	5.8	5.0	PL REC
TY	WP	2.4	2.4				2.4	18.4	1.3	1.3	2.7	27.0	−3.7	2.1	1.3	OTR
LY								16.2			1.7	21.2	1.7	2.1	1.3	PL REC
TY	WF	4.8	9.4				9.4	39.9	3.3	3.3	4.9	25.4	−5.0	3.7	2.2	OTR
LY		2.3	6.6				6.6	33.7	−.2	−.2	7.0	34.4	7.9	3.7	2.2	PL REC
TY	TC	6.6	11.3				11.3	28.1	.4	.4	4.4	33.5	−9.9	3.5	3.3	OTR
LY			2.2				2.2	33.8	.1	.1	2.3	20.1	4.8	3.5	3.3	PL REC
TY	KP	9.5	15.0				15.0	41.6	2.3	2.3	11.1	50.6	−5.4	6.4	5.0	OTR
LY		7.6	12.9				12.9	81.3			11.0	42.9	11.7	6.4	5.0	PL REC
TY	WG	2.4	2.9				2.9	30.0	2.5	2.5	6.6	64.6	−.4	3.8	1.1	OTR
LY								12.8			.7	8.6	1.0	3.8	1.1	PL REC
TY	FA			3.0			3.0	3.0			.8	70.2	12.3	7.8	10.8	OTR
LY		1.7	4.6				4.6	30.2	.2	.2	3.2	28.5	12.3	10.8	10.8	PL REC
TY	MI	7.7	13.5				13.5	56.9	3.6	3.6	5.4	24.9	−12.8	4.9	5.8	OTR
LY													6.3	4.9	5.8	PL REC
TY	RTV							7.6	−10.0	−10.0	−10.0		2.4			OTR
LY																PL REC
TY																OTR
LY																PL REC
TY																OTR
LY																PL REC
TY																OTR
LY																PL REC
TY																OTR
LY																PL REC
TY																OTR
LY																PL REC
TY																OTR
LY																PL REC
TY																OTR
LY																PL REC
TY																OTR
LY																PL REC
TY																
LY																
TY	EB	96.1	137.5	27.5			165.0	549.3	17.3	17.3	66.7	34.7	−89.4	75.4	72.5	
LT		11.5	52.9				52.9	435.3	3.3	3.3	51.2	28.5	84.4	102.9	72.5	
TY	ES	133.1	214.5	54.5			268.9	779.9	25.4	25.4	94.4	29.6	−113.	88.8	100.8	
LT		28.7	170.2				170.2	842.2	3.6	3.6	94.5	30.0	149.5	143.3	100.8	
TY	T	140.8	228.0	57.5			285.5	847.5	18.9	18.9	90.7	26.5	−111.	101.5	117.4	
LY		30.3	174.8				174.8	872.4	3.8	3.8	97.7	29.9	168.1	159.0	117.4	
TY	LST	152.2	347.6				347.6	918.7	.6							
LY	WK	32.4	205.4				205.4	897.6	9.1							
TY	2WK	26.4	154.6	202.9			357.5	979.9	1.5							
LY	AGO	4.6	168.8	139.9			308.8	975.7	16.2							
TY	3WK	32.6.	218.2	206.8			425.0	1,051	1.7							
LY	AGO	11.4	261.0	115.4			376.5	1,023	6.7							
TY	EOM	7,296	51,16	28,82			79,98	79,98								
LY	AUG	98.1	456.2	111.6			567.8	567.8								
TY	EOM	24	321.5	347.2			816.3	816.3								
LY	JUL	267.8	535.4	247.3			869.1	869.1								
TY	IMP															
LY	MEM	3														

The function of an open-to-buy report is to determine the dollar amount still available to purchase future merchandise. It is a control of the merchandising factors planned and identifies deviations between the planned goals and the actual results so that necessary corrective measures can be taken. Generally, it is prepared on a weekly basis. See pp. 304-307 for explanation of annotations.

3. Explanation of OPEN-TO-BUY REPORT (Figure 41)

Figure 41 illustrates a typical OTB report in regard to information that is contained in this document, (e.g., sales, stocks, markdowns, outstanding orders, and open-to-receive). The report format and frequency of issuance can vary with organizational size and needs. This particular report shown gives the reviewer the opportunity to analyze the department's performance by individual store or on a combined all-store basis. The annotated information on Figure 41 is numbered (1)-(9), and the following information corresponds to those numbers:

(1)   Date — Published date of the report, example: 10/10/93.

(2)   WEEK 01 OF 04 — Week of the month being viewed.

(3)   DEPT 736 BRIDGE SUITS — Department number and name.

(4)   Branch store listing — Listing of all the stores by location, example NY (left-hand column). At the bottom of the sheet is the total (T) section, which designates the total for all stores.

(5)   SALES — Net sales, by store, and the % change, which indicates the variance of CURRENT sales figures from PL and LY's actual. This category is divided into the following sections:

(a) CURRENT WK — Sales in $ and % by store, for the current week, example:

For store NY, CURRENT WEEK:

	$	%
TY —	$18.6	
PL —	$14.4	29%
LY —	$22.2	−16%

These figures show that for the current week, store NY had net sales of $18.6, which was 29% ahead of PL, but 16% behind LY.

(b) LAST WEEK — Sales in $ and % by store, for the prior week, example:

For Store NY, LAST WEEK:

	$	%
TY —	$9.7	
PL —	$14.2	−32%
LY —	$24.2	−60%

These figures show that for LAST WEEK, store NY had net sales of $9.7, which was 32% under PL, and 60% behind LY.

(c) 2 WKS AGO — Sales in $ and % by store, for 2 weeks prior, example:

For store NY, 2 WKS AGO:

	$	%
TY —	$23.3	
PL —	$14.2	64%
LY —	$17.0	37%

These figures show that for 2 WKS AGO , store NY had net sales of $23.3, which was 64% ahead of PL, and 37% over LY.

(d) PRIOR 4 WKS — Total sales in $ and % by store, for 4 weeks prior, example:

For store NY, PRIOR 4 WKS:

	$	%
TY —	$24.2	
PL —	$28.5	–15%
LY —	$26.5	–9%

These figures show that for PRIOR 4 WKS, store NY had net sales of $24.2, which was 15% under PL, and 9% behind LY.

(e) MTD (Month to date) — Total sales in $ and % by store, for fiscal month, example:

For store NM, MTD:

	$	%
TY —	$3.8	
PL —	$4.3	–11%
LY —	$8.5	–55%

These figures show that for MTD, Store NM had net sales of $3.8, which was 11% under PL, and 55% behind LY.

(f) STD (Season to date) — Total net sales in $ and % by store, from beginning of season to the current date, example:

For store BR, STD:

	$	%
TY —	$13.1	
PL —	$5.4	145%
LY —	$15.8	–17%

These figures show that for MTD, store BR had net sales of $13.1, which was 145% ahead of PL, but 17% behind LY.

(g) OCT $, NOV $, DEC $ — Planned sales in dollars for the present month and the next two months, compared to the same months LY, example:

For store BC:

OCT	NOV	DEC
$6.0 PL	$6.6 PL	$8.0 PL
$1.8 LY	$1.8 LY	$3.3 LY

(6) STOCK — Retail dollar value of stock currently received, which is planned by the month. This category is divided into the following sections:

(a) LAST WK. — Dollar value of stock LAST WK. compared to PL and LY amounts for the same week, example:

For store NY LAST WK.:

	$
TY —	$141.5
PL —	$177.3
LY —	$272.0

(b) CURRENT WK. — Current week's dollar value of stock compared to PL and LY amounts for the same week, example:

For store NY CURRENT WK.:

TY — $126.7

PL — $166.4

LY — $289.7

(c) OCT $, NOV $, DEC $ — Comparison of the PL EOM with LY's stock dollars for the current month, and the next two months, example:

For store NY:

OCT	NOV	DEC
$166.4 PL	$142.1 PL	$99.3 PL
$270.4 LY	$283.9 LY	$305.3 LY

(d) WKS. OF STOCK — This subcategory is divided into two sections:

(i) WKS. OF STOCK ON PL — Number of weeks merchandise should last based on planned sales. They indicate how many future weeks of planned sales will be covered by the current week's stock, example:

For store NY — 11 (weeks)

(ii) WKS. OF STOCK ON TRD — Weeks of stock adjusted for actual vs. planned sales performance, example:

For store NY — 8 (weeks)

(7)   ORDERS — Commitments by retail value of merchandise ordered. This category is divided into the following sections:

(a) PRIOR MOS. PAST DUE — Alerts the buyer to past due orders, TY vs. LY, example:

For store NY:

TY — $37.0

LY — $17.2

(b) OCT $, NOV $, DEC $ — Orders due for the current month, and the next two months, example:

For store NY:

OCT	NOV	DEC
$76.9 TY	$27.0 TY	
$117.2 LY		

(c) FUTURE — Figures for orders that would be due in future months, subsequent to the current month, and the next two months. Blank spaces indicate no future commitments.

(d) TOTAL — Figures that are the total of orders due for the present month, and next two months, example:

For store NY:

TY — $103.9

LY — $117.2

(e) STOCK AND ORDERS — Figures, which are a combination of the stock with the total orders, used to compare TY figures with LY figures, example:

> For store NY:
>
> TY — $230.7
>
> LY — $406.9

(8) MARKDOWNS — Dollar reductions taken. This category is divided into the following sections:

(a) CURRENT WEEK — Markdowns for the current week, comparing TY with LY, example:

> For store NY:
>
> TY — $8.0
>
> LY — $ .3

(b) MTD $ — Total dollar reductions taken month-to-date, comparing TY with LY, example:

> For store NY:
>
> TY — $8.0
>
> LY — $ .3

(c) STD $ — Total dollar reductions taken season-to-date, comparing TY with LY, example:

> For store NY:
>
> TY — $27.7
>
> LY — $43.3

(d) STD % SLS — Total reductions expressed as percentages of sales, comparing TY with LY, example:

> For store NY:
>
> TY — 21.9%
>
> LY — 32.0%

(9) OPEN TO REC. — Comparison of the planned and actual receipts at a given point in time during the month. It lists by month the retail dollar amount of merchandise open to be bought and received and is calculated by subtracting the month's orders from the month's planned receipts (i.e., planned purchases) figure. The positive numbers indicate money still available while the negative numbers show an over-extended or overbought condition. This overbought condition occurs when the month's orders placed exceed that month's planned receipts, example:

For store NY:

OCT	NOV	DEC
–$23.7 TY	$13.4 TY	$28.3 TY
$65.1 LY	$40.4 LY	$28.3 LY

FIGURE 42 Open-To-Buy Report

10/10/96 Week 01 of 04   Fall 1996   Open to Buy   Dept 862   Junior Denim Separates

	Cat	Prior 4 Wks	%	2 Wks Ago	%	Last Week	%	Current Week	%	MTD	%	STD	%	OCT $	NOV $	DEC $	Endin July $	Endin Aug $	Last Wk	Current Wk	OCT $	NOV $	DEC $	Wks Stk On Pl	On Trd
TY		44.7		17.7		14.4		16.4		16.4		183.8							140.0	151.6				11	7
PL	NY	20.7	116	10.4	70	10.4	39	11.1	48	11.1	48	107.6	71	44.1	59.1	72.1			157.8	168.6	168.6	177.5	144.5		
LY		19.7	128	10.8	64	8.9	63	10.9	50	10.9	50	103.5	78	42.9	42.9	68.6			142.7	153.7	133.6	170.6	259.5	11	
TY		3.4		1.5		1.5		1.7		1.7		31.7							28.0	34.6				12	9
PL	BR	2.2	53	1.1	35	1.1	33	1.2	45	1.2	45	19.8	60	5.1	1.30	17.8			28.9	51.3	51.3	58.6	43.8		
LY		2.2	51	1.0	52	1.0	44	1.5	14	1.5	14	19.5	63	4.7	4.7	16.6			21.7	21.6	22.4	36.6	51.4	9	
TY		3.8		1.6		1.3		2.0		2.0		31.0							27.8	34.6				11	17
PL	BC	5.5	−31	2.8	−41	2.8	−55	2.5	−19	2.5	−19	29.6	5	10.0	11.2	18.9			49.6	46.8	46.8	61.3	66.1		
LY		5.5	−31	2.2	−27	2.7	−53	2.1	−5	2.1	−5	28.1	10	9.7	9.7	17.9			51.9	59.7	66.6	78.9	64.2	19	
TY		2.6		1.0		1.6		1.7		1.7		21.2							19.9	25.5				13	10
PL	SH	1.9	38	.9	7	.9	81	1.6	9	1.6	9	10.5	102	6.4	7.0	11.4			36.5	33.3	33.3	42.1	34.6		
LY		1.7	54	.6	54	1.3	27	1.0	71	1.0	71	9.7	119	6.2	6.2	10.5			37.2	35.5	30.3	37.3	31.9	21	
TY		7.2		3.9		2.0		2.4		2.4		40.7							38.5	48.7				16	11
PL	NM	4.3	67	2.1	88	2.1	−5	2.1	15	2.1	15	25.9	57	8.4	14.4	19.8			43.4	56.0	56.0	64.1	46.9		
LY		3.4	110	2.3	72	1.6	23	1.7	38	1.7	38	24.4	67	8.1	8.1	18.5			43.2	51.1	54.9	54.9	63.0	19	
TY		9.2		4.1		4.0		2.8		2.8		41.3							49.9	54.8				15	10
PL	GC	5.1	81	2.5	64	2.5	58	2.6	8	2.6	8	27.1	53	10.4	12.8	25.3			51.7	50.7	50.7	77.8	61.5		
LY		4.1	123	2.8	49	2.4	63	3.2	−13	3.2	−13	26.6	55	10.1	10.1	23.8			41.9	49.1	52.9	70.2	69.1	14	
TY		1.7		.6		.3		.3		.3		13.7							17.8	21.9				27	22
PL	PG	.6	184	.3	93	.3	−15		31		31	5.8	136						8.3						
LY		.9	93	.1	897		119	.2		.2		6.0	128	.7	.7	7.3			3.6	3.3	4.4	16.0	16.5	7	
TY		2.9		1.1		1.2		.7		.7		16.3							17.2	20.6				28	25
PL	CH	2.5	17	1.2	−8	1.2	−1		17		17	11.0	48						20.7						
LY		1.5	94	.4	212	.7	65	.6		.6		8.3	95	2.8	2.8	5.9			23.0	22.6	23.3	27.4	20.9	22	
TY		9.7		4.1		2.7		3.2		3.2		48.1							62.4	73.1				15	18
PL	WP	9.4	3	4.7	−12	4.7	−42	3.6	−10	3.6	−10	48.6	−1	14.6	21.2	29.1			66.1	74.9	74.9	86.1	76.1		
LY		7.9	22	5.9	−30	3.6	−24	3.6	−10	3.6	−10	46.8	3	14.3	14.3	27.7			58.8	65.4	68.5	81.0	105.1	13	
TY		1.9		.9		1.0		.4		.4		15.7							16.4	19.2				9	21
PL	WF	4.3	−55	2.1	−56	1.7	−54	1.7	−75	1.7	−75	23.8	−34	6.8	9.2	11.6			37.9	40.9	40.9	43.0	43.0		
LY		4.0	−51	1.6	−43	2.0	−50	1.5	−72	1.5	−72	22.8	−31	6.6	6.6	11.0			33.3	38.1	40.3	42.8	47.6	19	
TY		4.8		2.9		2.2		2.2		2.2		34.3							36.5	41.1				18	15
PL	TC	4.3	11	2.1	36	2.1	7	1.7	31	1.7	31	25.6	34	6.5	9.8	12.8			36.5	42.4	42.4	47.6	43.8		
LY		3.8	26	2.1	38	1.8	25	2.0	14	2.0	14	25.4	35	6.4	6.4	12.4			38.0	35.5	38.8	46.6	51.4	17	
TY		2.2		1.2		.8		1.1		1.1		16.8							15.5	19.3				23	11
PL	KP	1.1	101	.5	142	.5	61	.6	90	.6	90	8.2	106	2.6	3.0	5.5			20.0	18.9	18.9	26.5	18.4		
LY		.8	189	.4	184	.5	53	.5	127	.5	127	7.9	113	2.6	2.6	5.2			15.0	16.0	15.6	23.4	18.1	22	
TY		2.3		1.3		1.0		1.1		1.1		17.1							19.6	24.9				16	20
PL	WG	2.8	−19	1.4	−11	1.4	−29	1.6	−33	1.6	−33	14.4	19	6.1	5.8	9.0			38.6	32.5	32.5	42.1	33.0		
LY		2.8	−19	1.6	−24	1.9	−48	1.4	−23	1.4	−23	16.4	5	6.6	6.6	10.4			22.5	28.3	28.8	39.3	38.3	17	
TY												20.5													
PL	FA	5.4	−100	2.7	−100	2.7	−100	3.5	−100	3.5	−100	36.0	−43	14.0	18.2	31.1			64.1	67.3	67.3	90.6	76.1		
LY		4.8	−100	1.7	−100	2.6	−100	2.5	−100	2.5	−100	33.4	−39	13.6	13.6	29.5			57.9	55.3	54.2	70.8	91.1	11	
TY		6.7		5.1		4.2		3.3		3.3		53.2							100.1	109.4				39	17
PL	MI	2.4	179	1.2	323	1.2	252	1.3	152	1.3	152	15.2	250	5.2	7.2	10.5			28.9	29.5	29.5	35.7	26.9		
LY																									
TY																			58.0	58.4				99	99
PL	RTV																								
LY																									
TY		51.7		24.2		19.5		19.7		19.7		328.0							349.4	418.5				16	16
PL	EB	44.0	17	21.7	12	21.7	−10	19.2	3	19.2	3	250.2	31	76.9	107.4	161.2			438.2	447.7	447.7	549.2	567.2		
LY		38.6	34	21.0	15	19.6	−1	19.4	2	19.4	2	241.8	36	78.9	78.9	167.0			390.0	426.4	447.0	554.3	577.5	16	
TY		96.4		41.9		33.9		36.1		36.1		511.8							489.4	570.1				14	11
PL	ES	64.7	49	32.1	31.	32.1	6	30.3	19	30.3	19	357.9	43	121.0	166.5	233.3			596.0	616.3	616.3	726.7	611.7		
LY		58.2	66	31.8	32	28.5	19	30.3	19	30.3	19	345.3	48	121.8	121.8	235.6			532.7	580.0	580.6	724.9	837.0	15	
TY		103.1		47.0		39.2		39.4		39.4		585.5							647.5	737.9				16	13
PL	T	72.5	42	36.0	31	36.0	6	35.1	12	35.1	12	409.1	43	140.2	191.9	274.9			689.0	713.1	713.1	853.0	714.7		
LY		63.0	64	33.5	40	31.1	23	32.8	20	32.8	20	378.8	55	135.4	135.4	265.1			590.6	635.3	634.7	795.7	928.1	14	

54. From the sample open-to-buy report, Figure 42 (on facing page left), from DEPT 862, JUNIOR DENIM SEPARATES, WEEK OF 10/10, for the FALL season, analyze the following:

(a.) What was the total dollar sales for DEPT 862? What was the percentage of variance from PL? From LY?

(b.) Which store reported no sales this year for PRIOR 4 WKS, 2 WKS AGO, LAST WEEK, or CURRENT WK?

(c.) Season-to-date (STD), which store had the largest sales increase? What was this increase in percent?

(d.) Season-to-date (STD), which store had the second largest sales this year? What was this increase in dollars?

(e.) What is the variance in percentage of DEPT 862 from PL? What is the variance in percentage from LY actual performance?

(f.) Calculate the sales percent increase planned for the combined sales of OCT $, NOV $, DEC $ over LY's performance for this period. On the basis of this department's STD sales performance, should the projected sales be revised? Why or why not?

(g.) What is the STOCK figure for the CURRENT WEEK TY for the total department? What was the PL figure?

(h.) Examine the performance of the Store WF. In light of its sales performance STD, what, if any, WK. OF STOCK (supply), on TRD would you advise?

FIGURE 43 Open-To-Buy Report

10/10 Week 01 of 04 Fall 1996 Open to Buy Dept 369 Junior Collections

		Orders							Markdowns				Open to Rec.			
		PRIOR MOS. PAST DUE	OCT $	NOV $	DEC $	FUTURE $	TOTAL $	S & O	CURRENT WEEK	MTD	STD	STD % SLS	OCT $	NOV $	DEC $	
TY	NY		57.7				57.7	157.5	.8	.8	31.3	35.8	−7.9	83.7	110.7	OTR
LY		101.9	141.1				141.1	391.7	10.4	10.4	49.3	25.7	62.9	83.7	110.7	PL REC
TY	BR	.6	19.6				19.6	64.5	.1	.1	10.3	41.3	−1.3	40.8	40.7	OTR
LY		18.8	18.8				18.8	59.8	1.8	1.8	7.8	19.5	18.1	40.8	40.7	PL REC
TY	BC	.1	17.3				17.3	49.7			10.0	37.6	−7.5	24.6	30.6	OTR
LY		20.1	22.7				22.7	73.9	2.9	2.9	15.0	36.7	13.9	24.6	30.6	PL REC
TY	SH	.1	11.7				11.7	37.3			7.9	45.7	−4.9	16.2	8.3	OTR
LY		15.8	19.8				19.8	48.5	1.8	1.8	11.9	44.1	9.0	16.2	8.3	PL REC
TY	NM	.1	26.3				26.3	60.3			10.7	38.0	3.3	24.3	41.3	OTR
LY		40.6	47.8				47.8	137.1	1.4	1.4	6.7	11.4	34.6	24.3	41.3	PL REC
TY	GC	.1	21.2				21.2	51.6	.1	.1	9.7	31.5	−3.7	49.6	31.0	OTR
LY		30.3	32.9				32.9	98.2	.4	.4	10.8	19.9	22.5	49.6	31.0	PL REC
TY	PG		11.6				11.6	29.5	.1	.1	12.6	79.6	−3.1	12.6	10.1	OTR
LY		9.1	9.1				9.1	21.4	.9	.9	1.3	20.0	8.5	12.6	10.1	PL REC
TY	CH	.1	16.8				16.8	46.4	−.1	−.1	7.5	47.6	−17.4	20.2	14.4	OTR
L		17.4	20.0				20.0	76.3	.9	.9	15.6	46.3	2.4	20.2	14.4	PL REC
TY	WP	.1	26.3				26.3	84.7			12.2	31.5	3.7	40.9	30.1	OTR
LY		32.3	44.2				44.2	120.9			19.4	30.5	35.0	40.9	30.1	PL REC
TY	WF	.1	14.0				14.0	40.6			9.2	48.9	−9.2	23.5	28.3	OTR
LY		25.4	31.1				31.1	80.2			10.5	34.1	8.9	23.5	28.3	PL REC
TY	TC	.1	15.3				15.3	46.7			9.0	37.4	−6.7	19.6	14.5	OTR
LY		18.9	21.5				21.5	78.4			7.7	25.4	13.1	19.6	14.5	PL REC
TY	KP		7.6				7.6	27.1	.1	.1	10.1	67.9	−3.4	8.3	2.4	OTR
LY		14.6	14.6				14.6	39.2	2.8	2.8	6.5	53.5	4.2	8.3	2.4	PL REC
TY	WG		7.6				7.6	21.8	.3	.3	6.7	50.5	−1.9	12.7	8.4	OTR
LY		12.6	12.6				12.6	49.8	2.1	2.1	10.3	72.1	5.7	12.7	8.4	PL REC
TY	FA	1.7	1.7				1.7	−1.8			5.1	59.3	6.7	43.1	30.0	OTR
LY		17.3	17.3				17.3	46.0	1.1	1.1	9.8	23.8	8.4	43.1	30.0	PL REC
TY	MI		15.8				15.8	48.2	.1	.1	3.2	21.8	−10.0	9.2	9.3	OTR
LY													9.4	9.2	9.3	PL REC
TY	RTV							65.7			−37.0		−.4			OTR
LY																PL REC
TY																OTR
LY																PL REC
TY																OTR
LY																PL REC
TY																OTR
LY																PL REC
TY																OTR
LY																PL REC
TY																OTR
LY																PL REC
TY																OTR
LY																PL REC
TY																OTR
LY																PL REC
TY																
LY																
TY	EB	1.4	195.5				195.5	560.3	.8	.8	111.5	42.3	−52.1	293.3	260.1	
LT		256.1	295.2				295.2	883.8	14.9	14.9	123.5	30.0	175.9	293.3	260.1	
TY	ES	1.4	253.2				253.2	717.7	1.6	1.6	142.8	40.7	−59.9	377.0	370.8	
LT		357.9	436.3				436.3	1,276	25.3	25.3	172.7	28.6	238.8	377.0	370.8	
TY	T	3.1	270.8				270.8	829.8	1.7	1.7	114.1	30.5	−63.6	429.3	410.1	
LY		375.2	453.6				453.6	1,322	26.3	26.3	182.5	28.3	256.6	429.3	410.1	
TY	LST	74.9	410.2				410.2	959.0	28.4	28.4						
LY	WK	60.4	60.4				60.4	963.0	1.2	1.2						
TY	2WK	29.6	75.6	262.0			337.6	976.1	5.7	5.7						
LY	AGO	18.2	76.6				76.6	1,037	4.5	4.5						
TY	3WK	29.6	141.1				141.1	751.6	−18.5	−18.5						
LY	AGO	71.8	235.2				235.2	1,141	−7.0	−7.0						
TY	EOM	4,777	41,72				41.72	41.72								
LY	AUG	347.1	456.0				456.0	456.0								
TY	EOM	30.1	289.0	23.4			312.4	312.4								
LY	JUL	124.3	124.3				124.3	124.3								
TY	IMP															
LY	MEM															

55. From the sample open-to-buy report, Figure 43 (on facing page left), from DEPT 369, JUNIOR COLLECTIONS, WEEK 10/10, for the FALL season, analyze the following:

(a.) What is the amount of overdue orders for DEPT 369, JUNIOR COLLECTIONS?

(b.) What are TY total outstanding orders in dollars? What were LY total outstanding orders in dollars?

(c.) What is the total department's STOCK AND ORDERS TY in dollars? Is this higher or lower than LY for this same period? By how much?

(d.) What amount of markdowns were taken in dollars STD TY? In percent?

(e.) For this year, which month has the largest scheduled OPEN-TO-RECEIVE for the department?

C. Operating Statements

Large stores periodically prepare a statement for each merchandise department, which itemizes and recapitulates the relationships of all factors affecting profit and/or loss. The report that reveals the profit position of a department is called an OPERATING STATEMENT. Operating statements are generally prepared monthly for each department and then combined into a final statement. The monthly statements are primarily designed to furnish gross margin information as well as other statistics that permit a careful analysis and that act as a guide for improving performance. This technique permits the buyer to evaluate and improve the operation of a department because it can: 1) Show the relationships of profit factors; 2) Reveal the effects on profits caused by changes; and 3) Permit comparison with other departments and other stores by means of percentages.

The terms "operating statement" and "profit and loss statement" are often used interchangeably. In reality, the operating statement provides an analytical statement that is useful in making changes if needed, and the final profit and loss statement is truly the "bottom line" that reflects the end results. These reports are generally issued at the end of a season or a year. While the use of computers permits ongoing operating statements, the form of these statements will vary from store to store, as does the amount of information that is presented. Figure 44, MERCHANDISE OPERATING STATEMENT, illustrates typical components of an operating statement.

FIGURE 44 Merchandise Operating Statement

		4 WEEKS ENDED				5 WEEKS ENDED			
		THIS YEAR		LAST YEAR		THIS YEAR		LAST YEAR	
		AMOUNT	%	AMOUNT	%	AMOUNT	%	AMOUNT	%
(1.)	# TRANSACTIONS & AVG. SALE								
(2.)	CUSTOMERS RETURNS & %								
(3.)	% CASH DISC. TO COST PURCH.								
(4.)	RETAIL PURCHASES & MU%								
(5.)	RETAIL STOCK END PERIOD & MU%								
(6.)	MARKDOWNS								
(7.)	SHORTAGES								
(8.)	EMPLOYEES' DISCOUNT								
(9.)	NET SALES PERIOD & % INCREASE								
(10.)	NET SALES-YEAR TO DATE & % INC.								
(11.)	ALTERATION COST								
(12.)	OTHER COST OF SALES								
(13.)	GROSS MARGIN & % TO SALES								
(14.)	CASH DISCOUNT & % TO COST								
(15.)	GROSS MARGIN PLUS DISCOUNT-PERIOD								
(16.)	GROSS MARGIN PLUS DISCOUNT-YEAR								
(17.)	MERCHANDISE SALARIES & EXPENSE								
(18.)	SALESPEOPLES' SALARIES & COMM.								
(19.)	STOCK & MISC. DIRECT SALARIES								
(20.)	PAYROLL TAX & EMPLOYEE BENEFITS								
(21.)	NEWSPAPER ADVERTISING								
(22.)	DIRECT MAIL & MAGAZINE ADV.								
(23.)	ADVERTISING PREPARATION								
(24.)	WINDOWS AND SIGNS								
(25.)	WRAPPING								
(26.)	DELIVERY								
(27.)	MERCHANDISE ADJUSTMENTS								
(28.)	SELLING SUPPLIES, ETC.								
(29.)	RENT								
(30.)	OCCUPANCY & HOUSEKEEPING								
(31.)	INTEREST & INSURANCE ON MERCH.								
(32.)	TOTAL CONTROLLABLE EXPENSES								
(33.)	DEPARTMENT CONTRIBUTION								
(34.)	INDIRECT EXPENSE								
(35.)	TOTAL EXPENSE								
(36.)	OPERATING PROFIT-PERIOD								
(37.)	OPERATING PROFIT-YEAR TO DATE								

The operating statement provides an analytical statement that is useful to evaluate and improve operations within a department. See pp. 314-315 for explanation of annotations.

1. Explanation of OPERATING STATEMENT (Figure 44)

To understand this report, it is necessary to analyze each item listed and then interpret the results. It is possible to compare a current merchandise operating statement with last year's statement for the same period as a means of measuring changes in performance, either for a specific single factor or the total picture. Additionally, there are several annual statistical reports (expressed in percentages) used by retailers for purposes of comparing operations. The best known is the Merchandising and Operating Results (MOR), published annually by the National Retail Federation, to which retailers contribute figures that are then collated and published. The annotated information on Figure 44 is numbered (1)-(37), and the following information corresponds to those numbers:

(1) # TRANSACTIONS & AVG. SALE — Number of customers the department is servicing in a given period, shows the actual dollar amount of the average transaction in the department, and reveals trends in price levels of the department.

(2) CUSTOMER RETURNS & % — Percentage of customer returns to gross sales and measures the efficiency of retaining sales originally made.

(3) % CASH DISC. TO COST PURCH. — Amount of cash discounts earned based on the cost of goods, and measures ability to take advantage of terms. (Controlling this factor is important because it frequently represents the difference between profit and loss.)

(4) RETAIL PURCHASES AND MU % — Result or accumulation of all initial markons for period to date and helps achieve desired markup needed.

(5) RETAIL STOCK END PERIOD AND MU % — Quantity of stock on hand at a retail value and the MU%, and helps in achieving profitable MU% at end of period.

(6) MARKDOWNS — Amount of markdowns ($ and %) taken to date. An analysis of those amounts can:

  ▶ Help prevent an excessive percentage of markdowns.

  ▶ Facilitate the investigation of the reason(s) for the excess.

  ▶ Prevent the reduction of the final gross margin to a low level.

(7) SHORTAGES — Monthly estimated difference between the book and the actual physical inventory. This figure does not necessarily change monthly, but is included in the calculation of the final gross margin percentage. Although management is responsible for shortage prevention and control, shortages affect the buyer's achievement of profitability.

(8) EMPLOYEES' DISCOUNT — Accumulation of the amount of employee discounts given to date. This figure affects gross margin in the same manner as the other retail reductions.

(9) NET SALES PERIOD AND % INCREASE — Dollar sales realized for period and the percentage of change, while also indicating the trend of a department's business.

(10) NET SALES — YEAR TO DATE AND % INCREASE — Dollar sales realized up to date and as an aggregate figure for the period.

(11) ALTERATION COST — Expenses charged against the selling departments that reduce the gross margin, making it an important figure to monitor.

(12) OTHER COST OF SALES — Figure that accounts for inward transportation of goods. This figure affects the gross margin in the same way as markdowns.

(13) GROSS MARGIN AND % TO SALES — Amount resulting from subtracting the total cost of goods from net sales, which is also expressed as a percentage of net sales.

(14) CASH DISCOUNT AND % TO COST — Dollar amount and its percentage relationship to net sales of discounts earned and deducted from billed costs. Lower than accepted trade discounts should not be accepted.

(15) GROSS MARGIN PLUS DISCOUNT — PERIOD — Amount of gross margin for the period and includes the cash discount amounts.

(16) GROSS MARGIN PLUS DISCOUNT — YEAR — Amount of gross margin for the year and includes the cash discount amounts.

(17)-(31) DIRECT EXPENSES — Various direct expenses, listed by classification, which affect the sales volume. These are expressed as a percentage of net sales.

(32) TOTAL CONTROLLABLE EXPENSES — Sum total of all direct departmental expenses.

(33) DEPARTMENT CONTRIBUTION — Figure that is obtained by deducting the "direct" operating expenses from the gross margin. This is considered to be the direct responsibility of the buyer.

(34) INDIRECT EXPENSE — Expenses that continue to exist even if department is discontinued. This figure is not considered to be a direct responsibility of the buyer.

(35) TOTAL EXPENSE — Summary of all expenses of operating the department and is a combination of direct and indirect expenses.

(36) OPERATING PROFIT — PERIOD — Indicates the profit performance for a designated time within a year.

(37) OPERATING PROFIT — YEAR TO DATE — A figure that denotes what is left of sales income after deducting both cost of sales and operating expenses. It shows a comparison between TY and LY figures, which can indicate a trend. The cumulative profitability form the beginning of the year to the current date can be noted.

FIGURE 45 Combined Deparment Operating Statement

Combined Department Operating Statement

(1) FALL 1996 (2) 525 MEN'S ACCESSORIES

(3) NET SALES, PCT CHANG * (4) MARKDOWNS–ACTU * (5) INVENTORY SHTAGE–A * (6) WORKROOM COST–A * (7) GROSS MARGIN

	NET SALES $TY	PCT	$LY	PCT	MRKD $TY	PCT	$LY	PCT	SHTG $TY	PCT	$LY	PCT	WKRM $TY	PCT	$LY	PCT	GM $TY	PCT	$LY	PCT
NY	1,740	28.7	1,352	46.4	58.1	3.3	49.3	3.6	141.7	8.1	15.3	1.1	7.6	.4	.4	–	931.1	53.5	753.4	55.7
FM	146.6	12.7	130.1	32.0	8.1	5.5	3.5	2.7	11.5	7.8	6.2	4.7	–	–	–	–	77.8	53.1	70.9	54.5
ST	165.4	2.7	161.1	28.7	8.3	5.0	5.4	3.3	16.6	10.0	8.8	5.4	–	–	–	–	86.6	52.3	86.8	53.9
BC	314.6	56.0	201.6	5.7	36.0	11.4	8.6	4.3	-1.9	-.6	16.4	8.1	–	–	–	–	170.6	54.2	105.1	52.2
SH	305.0	67.1	182.5	2.0	16.3	5.3	8.6	4.7	14.6	4.8	4.5	2.5	–	–	.1	–	166.4	54.6	99.6	54.5
GC	174.5	15.2	151.5	16.3	6.5	3.7	6.6	4.4	6.6	3.8	15.2	10.0	–	–	–	–	97.3	55.8	77.6	51.2
WP	369.6	36.1	271.6	31.9	17.4	4.7	8.6	3.2	21.2	5.7	5.8	2.1	–	–	–	–	201.1	54.4	150.7	55.5
WG	151.5	16.1	130.4	20.7	9.9	6.5	5.4	4.1	4.0	2.7	6.0	4.6	–	–	–	–	83.3	55.0	70.2	53.8
TC	194.7	34.3	145.0	6.9	8.1	4.2	4.8	3.3	3.4	1.8	3.9	2.7	–	–	–	–	110.0	56.5	79.9	55.1
WF	205.2	25.5	163.4	7.8	11.5	5.6	3.5	2.1	9.1	4.4	4.5	2.7	–	–	–	–	112.0	54.6	90.8	55.6
CH	171.9	23.1	139.7	5.5	7.8	4.5	5.7	4.1	12.7	7.4	9.4	6.7	–	–	–	–	92.4	53.8	73.9	52.9
KP	207.5	58.9	130.6	24.5	9.9	4.8	5.5	4.2	6.7	3.2	7.6	5.8	–	–	–	–	115.2	55.5	69.5	53.2
VV	175.4	-.3	175.8	10.0	15.6	8.9	5.8	3.3	-6.9	-3.9	9.9	5.6	–	–	–	–	99.8	56.9	94.5	53.7
FA	201.0	-.2	201.5	16.7	17.7	8.8	14.1	7.0	.8	.4	6.1	3.0	–	–	–	–	110.5	55.0	107.3	53.2
BR	276.8	54.4	179.3	–	13.5	4.9	2.8	1.6	5.3	1.9	4.4	2.5	–	–	–	–	155.3	56.1	100.6	56.1
CC	490.7	12.8	435.0	86.2	18.0	3.7	12.5	2.9	-7.6	-1.5	-16.6	-3.8	–	–	–	–	276.4	56.3	252.1	57.9
CP	–	–	–	–	–	–	–	–	–	–	–	–	–	–	–	–	–	–	–	–
HP	–	–	–	–	–	–	–	–	–	–	–	–	–	–	–	–	–	–	–	–
T	5,290	27.4	4,151	36.0	262.5	5.0	150.8	3.6	237.9	4.5	107.3	2.6	7.7	.1	.7	–	2,886	54.6	2,283	55.0

(8)(a) NEWSPAPER LINAGE * (b) OTHER DIRECT ADVERTI * (c)(d) SUPV-BUYING PAYROLL * (e) SELLING PAYROLL * (f) STOCK-CLERICAL PAYROLL

	NEWS $TY	PCT	$LY	PCT	ODA $TY	PCT	$LY	PCT	SUPV $TY	PCT	$LY	PCT	SELL $TY	PCT	$LY	PCT	STCK $TY	PCT	$LY	PCT
NY	13.9	.8	7.0	.5	11.0	.6	12.7	.9	63.3	3.6	53.2	3.9	97.5	5.6	76.9	5.7	24.9	1.4	21.6	1.6
FM	.5	.3	.3	.2	1.3	.9	.9	.7	6.7	4.6	7.5	5.7	10.4	7.1	8.1	6.2	2.0	1.4	1.8	1.4
ST	1.0	.6	.7	.4	1.8	1.1	1.2	.7	7.8	4.7	6.6	4.1	12.9	7.8	10.8	6.7	1.6	1.0	2.0	1.2
BC	1.2	.4	.7	.4	2.2	.7	1.3	.7	9.6	3.1	6.4	3.2	21.7	6.9	17.1	8.5	3.5	1.1	3.4	1.7
SH	1.4	.5	.7	.4	2.2	.7	1.3	.7	8.7	2.9	6.3	3.4	22.7	7.5	11.4	6.3	2.9	1.0	1.0	.6
GC	1.2	.7	.8	.5	2.0	1.1	1.2	.8	6.4	3.7	6.1	4.0	13.4	7.7	12.8	8.5	1.7	.9	1.2	.8
WP	1.6	.4	.9	.3	2.6	.7	1.6	.6	13.3	3.6	9.3	3.4	22.7	6.1	18.7	6.9	1.9	.5	1.3	.5
WG	.9	.6	1.1	.8	1.0	.6	.9	.7	3.6	2.4	5.2	4.0	15.3	10.1	11.0	8.5	.3	.2	.1	.1
TC	.6	.3	4.5	3.1	1.3	.7	.9	.6	7.4	3.8	6.5	4.5	15.0	7.7	10.2	7.0	5.1	2.6	.8	.5
WF	.7	.3	5.5	3.4	1.6	.8	.9	.5	8.0	3.9	6.5	4.0	17.4	8.5	12.4	7.6	3.9	1.9	1.0	.6
CH	.6	.3	.9	.6	1.4	.8	1.1	.8	6.4	3.7	4.9	3.5	11.5	6.7	8.1	5.8	1.0	.6	.6	.5
KP	1.2	.6	1.0	.8	1.2	.6	.9	.7	7.8	3.8	5.2	4.0	11.3	5.5	9.2	7.1	1.4	.7	1.1	.9
VV	1.0	.6	3.9	2.2	1.7	.9	1.7	.9	8.9	5.1	8.1	4.6	18.6	10.6	15.6	8.9	2.1	1.2	1.8	1.0
FA	1.2	.6	6.3	3.1	1.6	.8	1.3	.6	6.7	3.4	8.2	4.0	17.2	8.6	16.2	8.1	3.9	1.9	1.8	.9
BR	2.4	.9	5.9	3.3	1.9	.7	1.4	.8	9.3	3.4	6.3	3.5	22.6	8.2	13.1	7.3	4.1	1.5	1.1	.6
CC	-.1	–	-.1	–	59.8	12.2	53.7	12.3	8.2	1.7	7.5	1.7	9.2	1.9	–	–	–	–	–	–
CP	–	–	–	–	–	–	–	–	–	–	–	–	–	–	–	–	–	–	–	–
HP	–	–	–	–	–	–	–	–	–	–	–	–	–	–	–	–	–	–	–	–
T	29.3	.6	40.0	1.0	94.5	1.8	82.7	2.0	182.4	3.4	153.9	3.7	339.3	6.4	251.6	6.1	60.3	1.1	40.6	1.0

(g) DELIVERY * (h) RENT * (i) ALL OTHER EXPENSES * (8) TOTAL EXPENSES * (9) PRETAX PROFITS

	DEL $TY	PCT	$LY	PCT	RENT $TY	PCT	$LY	PCT	AOE $TY	PCT	$LY	PCT	TOT $TY	PCT	$LY	PCT	PTX $TY	PCT	$LY	PCT
NY	.4	–	.1	–	49.2	2.8	47.6	3.5	310.4	17.8	358.1	26.5	570.5	32.8	577.6	42.7	360.7	20.7	175.7	13.0
FM	–	–	–	–	5.4	3.7	3.1	2.4	24.3	16.6	21.5	16.6	50.5	34.5	43.1	33.1	37.3	18.6	27.8	21.3
ST	-.1	-.1	–	–	4.0	2.4	5.0	3.1	25.5	15.4	24.0	14.9	54.6	33.0	50.2	31.2	32.0	19.3	36.6	22.7
BC	-.1	–	–	–	8.0	2.5	2.5	1.2	43.0	13.7	25.9	12.8	89.1	28.3	57.3	28.4	81.6	25.9	47.8	23.7
SH	–	–	-.1	–	3.9	1.3	1.2	.7	39.3	12.9	21.6	11.8	81.1	26.6	43.5	23.8	85.3	28.0	54.1	30.7
GC	–	–	-.1	–	5.1	2.9	5.0	3.3	27.6	15.8	23.7	15.6	57.3	32.8	50.8	33.6	40.0	22.9	26.7	17.7
WP	-.3	-.1	-.2	-.1	5.7	1.6	6.0	2.2	50.8	13.7	36.8	13.5	98.4	26.6	74.3	27.4	102.7	27.8	76.4	28.1
WG	–	–	–	–	3.4	2.2	.4	.3	25.1	16.5	21.4	16.4	49.6	32.7	40.1	30.7	33.7	22.3	30.1	23.1
TC	-.1	–	.1	–	3.4	1.8	3.0	2.1	29.6	15.2	21.8	15.0	62.4	32.0	47.6	32.9	47.6	24.4	32.3	22.3
WF	-.1	-.1	.1	-.1	8.6	4.2	3.5	2.1	31.5	15.3	24.0	14.7	71.5	34.8	53.6	32.8	40.6	19.8	37.2	22.8
CH	-.1	–	–	–	9.9	5.7	6.4	4.5	25.5	14.8	19.3	13.8	56.2	32.7	41.2	29.5	36.2	21.1	32.6	23.4
KP	–	–	–	–	4.6	2.2	2.5	1.9	31.9	15.4	19.2	14.7	59.5	28.6	39.2	30.0	55.8	26.9	30.3	23.2
VV	–	–	.1	.1	2.9	1.6	2.9	1.7	32.6	18.6	34.6	19.7	67.7	38.6	68.7	39.1	32.1	18.3	25.8	14.7
FA	.1	–	.2	.1	5.7	2.8	5.2	2.6	32.9	16.4	30.3	16.8	69.3	34.5	73.1	36.3	41.2	20.5	34.2	17.0
BR	–	–	-.2	-.1	7.9	2.8	5.3	3.0	48.4	17.5	31.5	17.5	96.6	34.9	64.3	35.9	58.7	21.2	36.3	20.2
CC	-11.5	-2.3	-9.0	-2.1	19.9	4.1	1.4	.3	134.2	27.3	103.6	23.8	219.7	44.8	157.0	36.1	56.7	11.5	95.0	21.8
CP	–	–	–	–	–	–	–	–	–	–	–	–	–	–	–	–	–	–	–	–
HP	–	–	–	–	–	–	–	–	–	–	–	–	–	–	–	–	–	–	–	–
T	-12.0	-.2	-8.9	-.2	147.5	2.8	100.9	2.4	912.5	17.2	820.8	19.8	1,754	33.2	1,482	35.7	1,132	21.4	800.9	19.3

This report is a variation on an operating statement. This particular form indicates the individual store performance within one department.

2. Explanation of COMBINED DEPARTMENT OPERATING STATEMENT (Figure 45)

Figure 45 illustrates a combined department operating statement. It reports and compares the individual stores of a department for TY vs. LY in regard to net sales, markdowns, shortages, gross margin, direct expenses, and pre-tax profits. The annotated information on the Figure 45 is numbered (1)-(9) and the information on pp. 317-318 corresponds to those numbers.

(1) FALL, 1993 — Operating results for season being viewed.

(2) 525 — MEN'S ACCESSORIES — Department number and name.

(3) NET SALES, PCT CHANGE — Net sales, by store and percent of change that indicates the variance of TY actual sales from LY actual sales, example:

For store NY:

	$	%
TY	$1740	
LY	$1352	28.7

(4) MARKDOWNS — ACTUAL — The dollar reductions taken TY and LY. The dollar amount of each year is expressed as a percentage of the net sales, example:

For store NY:

	$	%
TY	58.1	3.3
LY	49.3	3.6

(5) INVENTORY SHTAGE — ACTUAL — The difference between the amount of recorded inventory on hand and that which was counted. The dollar amount of shortage as well as the percent for TY and LY is reported and this shortage is calculated as a percentage of sales, example:

For store NY:

	$	%
TY	141.7	8.1
LY	15.3	1.1

(6) WORKROOM COST — Charges against a selling department for repairing or preparing goods for sales. It is expressed as a percent of net sales, example: For total department (T), the workroom cost of $7.7 for the entire department is .1%.

(7) GROSS MARGIN — The difference between the net sales and cost of goods is noted in dollars, for each store, and expressed as a percentage of sales for TY and LY, example:

For total department (T):

	$	%
TY	2886	54.6
LY	2283	55.0

(8) TOTAL EXPENSES — The sum total of all the direct expenses plus all other expenses are listed by store, in dollars, TY and LY, and are expressed as a percent of the net sales for each store, example:

For store NY:

	$	%
TY	570.5	32.8
LY	577.6	42.7

The TOTAL EXPENSES are classified further by function or activity and are listed (a)-(i). These annotations are explained on p. 318.

(a.) NEWSPAPER LINAGE — Indicates the cost of newspaper advertising.

(b.) OTHER DIRECT ADVERTISING — Could include magazine, bill enclosures, and other direct mail pieces.

(c.) SUPV — Supervisory salaries of various management levels.

(d.) BUYING PAYROLL — The salaries of the buyer, associates, and assistants who are involved in purchasing goods for resale.

(e.) SELLING PAYROLL — The salaries of sales personnel and any other selling methods used.

(f.) STOCK-CLERICAL PAYROLL — The personnel involved directly in keeping the stock in salable condition as well as its flow from receipt to sale.

(g.) DELIVERY — The cost incurred in shipping merchandise to the consumer.

(h.) RENT — The cost of maintaining space within the store.

(i.) ALL OTHER EXPENSES — This figure generally includes those cost of running the business that are grouped as "indirect".

(9) PRE-TAX PROFITS — This represents the resultant amount when income exceeds expenses before the payment of taxes. It is listed for TY and LY for store in the dollar amount and also is expressed as a percentage of sales, example:

For total department (T):

	$	%
TY	1132	21.4
LY	800.9	19.3

FIGURE 46 Departmental Operating Statement

Departmental Operating Statement					
FALL 1996 525 MEN'S ACCESSORIES					
	$ 1993 TY		$ 1992 LY	PCT	
(1) NET SALES– NY	1,740	28.7	1,352	46.4	
(2) FM	146.6	12.7	130.1	32.0	
(3) ST	165.4	2.7	161.1	28.7	
(4) BC	314.6	56.0	201.6	5.7	
(5) SH	305.0	67.1	182.5	2.0	
(6) GC	174.5	15.2	151.5	16.3	
(7) WP	369.6	36.1	271.6	31.9	
(8) WG	151.5	16.1	130.4	20.7	
(9) TC	194.7	34.3	145.0	6.9	
(10) WF	205.2	25.5	163.4	7.8	
(11) CH	171.9	23.1	139.7	5.5	
(12) KP	207.5	58.9	130.6	24.5	
(13) VV	175.4	–.3	175.8	10.0	
(14) FA	201.0	–.2	201.5	16.7	
(15) BR	276.8	54.4	179.3	–	
(16) CATALOG	490.7	12.8	435.0	86.2	
(17) CP	–	–	–	–	
(18) HP	–	–	–	–	
(19) COMBINED	5,290	27.4	4,151	36.0	
(20) AVG GROSS SALE & PCT RETURNS	40.30	8.8	34.84	8.3	
(21) RETAIL PURCH & CUM MARKON PCT	5,267	54.1	5,001	53.3	
(22) MARKDOWNS	–	–	–	–	
(23) EMPLOYEE DISCOUNT	13.4	.8	11.0	.8	
(24) INVENTORY SHORTAGE	–	–	–	–	
(25) WORKROOM COST	7.6	.4	.4	–	
(26) DISC EARNED, PCT TO SALES	91.9	5.3	70.0	5.2	
(27) DISC TAKEN, FRT, PCT COST PUR	10.5	.4	10.4	.2	
(28) GROSS MARGIN	931.1	53.5	753.4	55.7	
(29) NEWSPAPER LINAGE	13.9	.8	7.0	.5	
(30) ALL OTHER ADVERTISING	11.0	.6	12.7	.9	
(31) NEWSPAPER PREPARATION	8.1	.5	3.9	.3	
(32) MDSE SUPERVISION & TRAVEL	45.4	2.6	39.5	2.9	
(33) BUYING PAYROLL & TRAVEL	17.9	1.0	13.6	1.0	
(34) BUYING CLERICALS PAYROLL	3.6	.2	2.6	.2	
(35) SELLING PAYROLL & COMMISSION	97.5	5.6	76.9	5.7	
(36) STOCK & FITTING ROOM CHECKER	24.9	1.4	21.6	1.6	
(37) DELIVERY	.4	–	.6	–	
(38) WRAPPING & PACKING	16.4	.9	113.0	8.4	
(39) MARKING	7.9	.5	8.3	.6	
(40) MAIL & TELEPHONE ORDER	2.7	.2	.4	–	
(41) MERCHANDISE ADJUSTMENT	1.4	.1	1.8	.1	
(42) TOTAL DIRECT EXPENSE	251.0	14.4	301.8	22.3	
(43) INCOME AFTER DIRECT EXPENSE	680.1	39.1	451.6	33.4	
(44) HOUSEKEEPING	55.6	3.2	58.3	4.3	
(45) RENT-SELLING	49.2	2.8	47.6	3.5	
(46) RENT-NONSELLING	–	–	.2	–	
(47) INDIRECT EXPENSE	214.7	12.3	169.8	12.6	
(48) TOTAL GENERAL EXPENSE	319.5	18.4	275.8	20.4	
(49) PROFIT BEFORE TAXES	360.6	20.7	175.8	13.0	
(50) TURNOVER & PCT DEPT SALES TO	1.6	.8	2.0	–	
(51) TRANSACTION & PCT CHANGE	47.3	11.9	42.3	23.4	
(52) SELLING AREA & SALES/SQ FOOT	1859	936	2169	623	

BRANCH SALES PER SQUARE FOOT										
1993 TY	FM– 209	ST– 251	BC– 237	SH– 609	GC– 274	WP– 821	WG– 360	TC– 389	WF– 175	
1992 LY	273	208	486	1342	238	384	– 384	– 263	290	

1993 TY	187	KP– 299	VV– 596	FA– 321	BR– 282	CC–	MC–	–
1992 LY	206	265	598	372	184	–	–	–

This report shows the detailed performance for Store NY only.

3. Explanation of DEPARTMENTAL OPERATING STATEMENT (Figure 46)

Figure 46 itemizes in detail the factors affecting profit. An analysis of this report reveals the profit factors that may be changed or adjusted to improve the final results. The annotated information on Figure 46 is numbered (1)-(52), and the information on pp. 320-321 corresponds to those numbers.

(1)   NET SALES — Gross sales minus returns, allowances, and discounts for each store (01-18). The departmental sales are given for the current season (FALL) and the percent of change from LY as well as the net sales for the comparable season of LY. Also shown is the percent of change for the comparable season from the prior year. The last item on the NET SALES list is the COMBINED NET SALES (19) of all stores in dollars and percents.

(20)   AVG GROSS SALE & PCT RETURNS — Total sales (gross sales) before the deduction of returns and allowances. The dollar figures are calculated by dividing the gross sales by the total number of transactions. The percentage figures are a percent of the gross sales.

(21)   RETAIL PURCH & CUM MARKON PCT — Department's total purchases for the current season and comparable season of LY. This figure also records the cumulative markon percentage for the current season and the comparable season of last year.[8]

(22)   MARKDOWNS — Dollar amount of reductions in dollars and percentages. Also shown are the results of this season of LY.

(23)   EMPLOYEE DISCOUNT — All discounts that are allowed to employees combined. This is considered a retail reduction.

(24)   INVENTORY SHORTAGE — Figure that is calculated by an adjustment between the physical count and book stock. It is also included in the determination of the final gross margin percentage.

(25)   WORKROOM COST — Cost of making an item saleable in a department, which is included in the calculation of cost of goods sold.

(26)   DISCOUNT EARNED PCT TO SALES — Total store departmental average of cash discount, deducted from billed costs on purchases.

(27)   DISC TAKEN FRT PCT COST PUR — Actual discount dollars taken for the current season. Also shown is the freight percentage to cost purchases for the season.

(28)   GROSS MARGIN — Net sales minus the total cost of goods sold. Shown in dollars and percent for the current and comparable season LY.

(29)-(33) DIRECT EXPENSES — Are divided into the following categories:

(29) NEWSPAPER LINAGE — Cost based on the number of lines advertised.

(30) ALL OTHER ADVERTISING — Actual advertising costs of radio, television, bill inserts, catalogs, etc.

(31) NEWSPAPER PREPARATION — Administrative costs of the advertising and art departments based on linage.

(32) MDSE SUPERVISION & TRAVEL — Salary and travel expenses of each merchandise manager, which is charged directly to the selling departments that are supervised by those merchandise managers.

(33) BUYING PAYROLL & TRAVEL — Salary and travel expenses of each buyer, associate buyer, assistant buyer, and group manager charged directly to the assigned selling department. Expenses are prorated on the basis of net sales if the "executive" is assigned to more than one department.

---

[8]The remaining percentage figures in this report (Figure 46) are expressed as a percent to net sales unless otherwise stated.

(34) BUYING CLERICALS PAYROLL — Expenses for merchandise clericals.

(35) SELLING PAYROLL & COMMISSIONS — Salaries and commissions of sales personnel. These figures are generated from a weekly cost report.

(36) STOCK & FITTING ROOM CHECKERS — Salaries for stock personnel and fitting room checkers. This figure also includes the warehouse stock personnel payroll.

(37) DELIVERY — Actual expenses of customer delivery, which is charged to each department.

(38) WRAPPING AND PACKING — Figures derived by using wrapping rates for "takes" and "sends" that are developed periodically. Figured on the basis of the rates multiplied by the take and send transactions.

(39) MARKING — Actual expense assigned by cost of time of receiving and marking of merchandise.

(40) MAIL & TELEPHONE ORDER — Expense charged for the department's transaction handled in the mail and telephone order rooms.

(41) MERCHANDISE ADJUSTMENT — Expense that is calculated by an adjustment rate derived by dividing the total expense of the merchandise adjustment department by the number of adjustment units. This is charged directly to the department for the number of adjustments handled for that department for the season.

(42) TOTAL DIRECT EXPENSE — Total dollar amount of direct expenses for the season, derived from adding lines (29)-(33) of this report.

(43) INCOME AFTER DIRECT EXPENSE — Figure that is calculated by subtracting line (42) from line (28) and denotes the difference between the gross margin and total direct expense.

(44) HOUSEKEEPING — Cost of maintenance, cleaning, etc., which is prorated to all departments based mainly on space, with the remaining percent based on sales.

(45) RENT-SELLING — Rent charged to a department that is based on that department's square footage and location in the store.

(46) RENT-NONSELLING — Rent charged to a department for warehouse and reserve stockroom space.

(47) INDIRECT EXPENSES — Actual expenses that include receiving, supervision of selling floor, accounting, data processing, etc. This is allocated on sales and is based on various appropriate formulae.

(48) TOTAL GENERAL EXPENSE — Total of lines (44)-(47).

(49) PROFIT BEFORE TAXES — Difference between line (43) INCOME AFTER DIRECT EXPENSE, and line (48) TOTAL GENERAL EXPENSE.

(50) TURNOVER & PCT DEPT SALES TO TOT — Relationship between an average stock and sales. Derived by comparing the percent of a department's total sales volume to the volume of the total store.

(51) TRANSACTION AND PCT CHANGE — Number of sales transactions, which reflects the percent of increase or decrease from LY.

(52) SELLING AREA & SALES/SQ FOOT — Amount of departmental selling area in square feet as compared to LY. This category also shows sales per square foot, which is calculated by dividing the selling area in square footage into the net sales. The last set of figures — BRANCH SALES PER SQUARE FOOT — presents branch sales per square foot TY vs. LY. To be used for comparison between all the branches.

NOTES

► **Use Figure 45 for practice problems 56–63.**

56. What is the total net sales TY in dollars? What percent change does this represent? What was the total net sales LY in dollars?

57. What store generated the highest sales percent of change over LY? What percent is this variance?

58. What store achieved the largest dollar sales TY in the total department? What was the dollar amount of these sales? What percent of the total sales volume for DEPT 525 is this?

59. What store had the largest percent of markdowns this year? What was the amount of these markdowns in percentages? LY, this store's markdown percent was on a par with the other branches. What action, if any, should be taken?

60. What was store ST's inventory shortage TY (in percent)? What is the total departmental shortage (in percent)? The total departmental shortage TY is how much higher than LY? What preventive measures would you recommend to stop this trend?

61. What are the two greatest individual expenses listed this year?

62. How much lower are the total departmental expenses TY vs. LY (in percent)?

63. How much (in percent) did the pre-tax profit increase TY vs. LY?

▶ **Use Figure 46 for practice problems on this page.**

64. List and compare the critical major profits figures for this entire department.

65. What factors contributed to the improved profit? How? Why?

66. Which profit factors require improvement? Why?

Selected Answers

UNIT I

1. 5%
3. $49,122.00
5. $40,000.00
7. 1.5% (Junior Dept.); 4.5% (Misses Dept.)
11. $2,917.40
13. $59,045.00
15. (a.) $1,374.00; (b.) $1,420.26
19. $8,100.00; 2.9%
21. $225,500.00; 46%
23. $2,500.00
25. (a.) $9,000.00; (b.) 41.25%
27. $38,800.00; 48.5%
33. 51%
35. (a.) $261,000.00; (b.) $139,000.00
37. $608,000.00
41. (a.) $236,160.00; (b.) $238,040.00; (c.) $15,800.00; 3.2%
43. $5,620.00
45. $9,561.93
47. $350,000.00
49. 50.4%
51. $50,000.00
53. $25,000.00
55. 50.19%

UNIT II

1. $29.95
3. $93.00
5. 45%
7. (a.) $88.54; (b.) $90.00; (c.) 46.54%
11. $75.43; 50%
13. 50.0%
15. $22.00
17. 48.71%
19. $41.67
21. $23.07
23. (a.) $25.00; (b.) $10.83

25. $975.00
27. $1,225.00
29. $520.00
33. $669.75
35. (a.) $1,330.00; (b.) 15.34%
39. 3.8%
41. $20.57
43. $8.57
45. 47.95%
47. (a.) 47 %; (b.) 2.98% or 3%
48. 54.4%
50. $4,240.00
52. $67.90
54. (a.) $117,000.00; .5% More

UNIT III

1. 51.4%
3. 49.15%
5. 53%
7. 53.8%
9. 48.3%
11. 53.7%
13. (a.) 55.3%; (b.) 54.9%
15. 50.1%
17. 37.3%
21. 13.4%
23. (a.) $8,704.00; (b.) $108.84
25. $1,774.50
27. $27.41
28. $25.51
29. $26.00
31. 53.7%
33. 49.6%
35. 47.2%
37. 47.6%
38. 13%
40. (a.) 41.6%; (b.) 47.6%
42. 53.64%

45. $21.68
46. 61.54%
47. $5.83
48. $47.14
50. $60.41
53. 49.52%
55. 49.57%
58. (a.) $130,000.00; (b.) $139,200.00

UNIT IV

1. $62,875.00
3. $95,000.00
5. (a.) $23,000.00; (b.) $11,730.00
7. (a.) $1,311,000.00; (b.) $672,543.00
9. $162,382.00
11. (a.) $937,000.00; (b.) $475,996.00
14. .41%
16. 1.5%
18. 2.07%
20. $70,000.00
22. (a.) .1% Higher; (b.) $2,000.00 More
27. (a.) $284,100.00; (b.) .$282,550.80

UNIT V

1. 10%
3. August—$900,000.00
   September—$840,000.00
   October—$960,000.00
   November—$1,080,000.00
   December—$1,500,00.00
   January—$720,000.00
5. 20%
7. $923,400.00
9. 3.88 or 3.9
11. $80,000.00
13. $300,000.00
15. 2.5
17. (a.) $40,000.00; (b.) 4.95
19. 3.2
21. 1.10
23. 3.0
25. $312,000.00
27. 4.25

29. $117,000.00
31. $126,150.00
33. $275,000.00
35. $60,000.00
37. $80,000.00
39. (a.) $20,000.00; (b.) $9,700.00; (c.) .36
41. (a.) $28,000.00; (b.) $33,000.00
43. $61,800.00
45. None
47. None
49. None
51. 10.4
53. $780,000.00
55. 10.2%
57. (a.) $3,000.00; (b.) $1,320.00
59. $37,000.00
61. 1.5
63. (a.) 4.5; (b.) 2.1
65. 3.58
66. August—$217,250.00 BOM
    September—$230,050.00 BOM
    October—$266,550.00 BOM
    November—$233,450.00 BOM
    December—$234,750.00 BOM
    January—$216,550.00 BOM
68. $88,000.00

UNIT VI

1. $122.50
5. $972.00
7. $488.88
9. $4,419.80
11. $4,159.67
13. $522.00
15. $624.00
17. May 11
19. B.
21. $5,320.00
27. $4,443.75
29. $831.25
31. $2,047.50
33. $2,437.50
35. $6,446.86

36. December 5

38. (a.) March 10; (b.) March 30; (c.) $5,674.39

40. $2,928.00

45. $1,404.00

UNIT VII

3. (a.) GC Store, 126%; (b.) $141.2; (c.) NY Store, $40.5; (d.) WG Store, $3.8; (e.) GC Store 447%

11. $15.01-$20.00

13. Stores 01 and 05

21. 82, BRUSHED DENIM JEANS

23. $15.0

27. $17,814.00 Estimated April sales

33. Styles #7550, #7552, and #7557; Style #7552 Requires immediate shipment for delivery.

37. Dept. 220—Belts

43. Week ending Feb. 19, with sales of $86.4; Week ending Mar. 12 had best Percent Increase.

45. Week ending Mar. 19 had the greatest dollar markdowns, which equalled $122.8.

55. (a.) $3.1; (b.) $270.8 Total O.O. TY; $453.6 Total O.O. LY (c.) $829.8; $492.2 Lower than LY; (d.) $114.1 or 30.5% Markdowns STD TY; (e.) November OTR

57. Store SH; 67.1% Variance

61. Selling payroll and supervisor-buying payroll.

63. 2.1% Profit increase TY over LY

Reference Readings

Berman, Barry, and Joel R. Evans, *Retail Management, fourth edition.* New York: MacMillan Publishing Co., 1989.

Bolen, William H., *Contemporary Retailing, third edition.* Englewood Cliffs, NJ: Prentice Hall, 1988.

Bohlinger, Maryanne Smith, *Merchandise Buying: A Practical Guide, third edition.* Boston: Allyn & Bacon, 1990.

Capron, H. L., and Brian K. Williams, *Computers and Data Processing, second edition.* Menlo Park, NJ: The Benjamin/Cummings Publishing Company, Inc., 1984.

Davidson, William R., Daniel J. Sweeney, and Ronald W. Stampfe, *Retailing Management, sixth edition.* New York: John Wiley & Sons, 1988.

Duncan, Delbert J., Stanley C. Hollander, and Ronald Savit, *Modern Retailing Management: Basic Concepts & Principles, tenth edition.* Homewood, IL: Richard D. Irwin, Inc., 1983.

Gillespie, Karen R., Joseph C. Hecht, and Carl F. Lebowitz, *Retail Business Management, third edition.* New York: Gregg Division, McGraw-Hill Book Co., 1983.

Pintel, Gerald, and Jay Diamond, *Retailing, fifth edition.* Englewood Cliffs, NJ: Prentice Hall, 1991.

Stair, Jr., Ralph M., *Principles of Data Processing, revised edition.* Homewood, IL: Richard D. Irwin, 1984.

Stone, Elaine, *Fashion Merchandising, fifth edition.* New York, Gregg Division of McGraw-Hill Book Co., 1990.

Wingate, John W., and Joseph Friedlander, *The Management of Retail Buying, second edition.* Englewood Cliffs, NJ: Prentice Hall, 1978.

Index

method of inventory, 132, 146, 155; and terms of sale, 226; and the use of computers, 265
Gross margin return for dollar of inventory (GMROI), 177; 265
Gross markdown, 74
Gross profit, 19, 23; and planning markups, 184; also see profit
Gross sales, 2, 23

H

Hardware, 261

I

Indirect expenses, 3, 7, 23
Initial markup, 54, 67, 69, 75, 76, 92; calculation of, 94, 95; and cumulative markups, 96; establishment of, 93; and maintained markups, 97; and planning markups, 184
Input, 261, 262
Input unit, 262
Invoice, 226-232
Inward freight, 6, 23, 253; and cumulative markups, 96; and initial markups, 92; and terms of sale, 226

J

Journal/purchase record, 138

L

List price, 226; and cash discounts, 228; and trade discounts, 226

M

Maintained markup, 97; impact on gross margin, 148
Markdown, 67, 71, 76; in calculating book inventory, 136, 138; in calculating initial markups, 94; and calculating open-to-buy, 205; and computer use in retailing, 262; and the dollar merchandise plan, 166; and open-to-buy reports, 205; and planned purchases, 185; planning, 183; in planning markups, 92, 184; POS markdowns, 138, 141; and pricing, 47; relationship to markup, 92; and turnover, 173
Markdown cancellation, 73
Markdown percentage, 67, 71; typical markdown percentages, 183 (Figure 28); also see markdown
Markon, see markup
Markup, 49, 53, 67; calculation of

initial markups, 94; cumulative markups, 96, 205; and the dollar merchandise plan, 166; establishing initial markups, 92; and GMROI, 177; initial markups, 92; limitations of markups percentage as a guide to profits, 119; maintained markups, 97; and planned purchases, 185; and planning markups, 184; relationship to profit, 92; and the retail method of inventory, 155; and trade discounts, 226; also see markup percentage
Markup cancellation, 75
Markup percentage, 53; in averaging markups, 109; in averaging retails, 111; cancellation of, 75; cumulative markup, 96; and the dollar merchandise plan, 166; and GMROI, 177; initial markup percentage, 93, 95; limitations of the markup percentage as a guide to profits, 119; maintained markup, 97; and planned purchases, 185; planning markups, 187; also see markup
Merchandise control, reports used in, 267, 273, 279, 283, 287
Merchandise transfer, see transfer of goods
Merchandising for a profit, 1
Merchandising and Operating Results of Department Specialty stores (MOR), 5; and planning monthly stock proportions, 179
Monitor, 262

N

Net cost (of merchandise sold), 21, 23, 226; and anticipation, 247; and discounts, 226, 229
Net markdown, 74
Net profit, see profit
Net sales, 2, 4, 9, 19, 23; and calculating book inventory, 136; and GMROI, 177; and gross margin, 97; and maintained markup, 97; and markdowns, 67, 70, 71; and planning markdowns, 183; and the retail method of inventory, 132; and shortages/overages, 153; and turnover, 173
Net terms, 232
Net period, 237, 240

Noncontrollable expenses, 3, 8; and pricing decisions, 47

O

Opening book inventory, see book inventory
Opening inventory, 21; at cost, 23; at retail, 23; in calculating cost of merchandise sold and gross margin, 146; and calculating open-to-buy, 207; and cumulative markup, 96; and planning markups, 184
Open-to-buy, 205; calculating at the beginning of a month, 207; calculating during the month, 207; calculating for balance of month based on planned closing stock, 208; calculating for balance of month based on predetermined planned purchases, 207; elements of report, 205; example of report, 206; and markdowns, 67; and quantity discounts, 228; and turnover, 173
Open-to-buy reports, 303, 308, 310
Open-to-receive, see planned purchases
Operating expenses, 7, 23
Operating statement, 313, 317, 320
Optical character recognition, 262
Ordinary dating, see regular dating
Original markup, 54, 67, 69, 75, 76; also see initial markup
Output, 260, 262
Overage, 133, 153; calculation of, 154; causes of, 153; when using the retail method of inventory, 155

P

Perpetual inventory, 76, 132; maintenance of, 135
Physical inventory, 133; in calculating book inventory, 136; and the retail method of inventory, 132, 155; and shortages/overages, 133, 153
Planned purchases, 185; converting retail purchases to cost, 186; and open-to-buy, 205; and the six-month seasonal dollar plan, 166
Planned receipts, see planned purchases